Cabinets of Experimental Philosophy in Eighteenth-Century Europe

History of Science and Medicine Library

VOLUME 40

Scientific Instruments and Collections

Studies Published under the Auspices of the Scientific Instrument Commission

VOLUME 3

The titles published in this series are listed at brill.com/hsml

Cabinets of Experimental Philosophy in Eighteenth-Century Europe

Edited by

Jim Bennett and Sofia Talas

BRILL

LEIDEN • BOSTON
2013

Cover illustration: Demonstration of an electrical plate machine in the Physics Theatre of *Felix Meritis*, probably by the lector Henricus Aeneae (Drawing by P. Barbiers and J. Kuyper, 1794, engraved by Reinier Vinkeles, 1801, detail).

Library of Congress Cataloging-in-Publication Data

Cabinets of experimental philosophy in eighteenth-century Europe / edited by Jim A. Bennett and Sofia Talas.
pages cm. — (History of science and medicine library, ISSN 1872-0684 ; volume 40) (Scientific instruments and collections ; volume 3)
Includes bibliographical references and index.
ISBN 978-90-04-25296-7 (hardback : acid-free paper) — ISBN 978-90-04-25297-4 (e-book)
1. Scientific apparatus and instruments—Collectors and collecting—Europe—History—18th century. 2. Cabinets of curiosities—Europe—History—18th century. 3. Science—Social aspects—Europe—History—18th century. 4. Physics—Social aspects—Europe—History—18th century. 5. Europe—Intellectual life—18th century. 6. Europe—Social conditions—18th century. I. Bennett, J. A. (James A.) II. Talas, Sofia.

Q185.C27 2013
507.2'04—dc23

2013014808

This publication has been typeset in the multilingual "Brill" typeface. With over 5,100 characters covering Latin, IPA, Greek, and Cyrillic, this typeface is especially suitable for use in the humanities. For more information, please see www.brill.com/brill-typeface.

ISSN 1872-0684
ISBN 978-90-04-25296-7 (hardback)
ISBN 978-90-04-25297-4 (e-book)

This book is printed on acid-free paper.

CONTENTS

LIST OF ILLUSTRATIONS

Colour plates can be found in a separate section following page xxiv.

NOTES ON CONTRIBUTORS

JIM BENNETT is Visiting Keeper at the Science Museum, London. He was formerly Director of the Museum of the History of Science and Professor of the History of Science in the University of Oxford. He is a previous President of the Scientific Instrument Commission.

PAOLA BERTUCCI is an Assistant Professor of History at Yale University. She has published on the history of eighteenth-century electricity and its medical applications, and is the author of *Viaggio nel paese delle meraviglie. Scienza e curiosità nell'Italia del Settecento* (Turin, 2007). She is now working on a book on science and secrecy in Enlightenment France.

PAOLO BRENNI specialises in the history of scientific instruments and of precision industry in the period from the beginning of the eighteenth to the mid-twentieth century. He is researcher for the Italian CNR (National Research Council), works for the Fondazione Scienza e Tecnica, Florence, and collaborates with the Museo Galileo, Florence. He is President of the Scientific Instrument Commission of the International Union of History and Philosophy of Science (IUHPS) and of the Scientific Instrument Society. In 2009 he was elected first vice-president of the Division of History of Science and Technology of the IUHPS.

TIEMEN COCQUYT is a curator at the Museum Boerhaave, the Dutch national museum for the history of science and medicine. He studied physics and history of science at Utrecht University, and previously was Assistant Curator at Utrecht University Museum. Research interests include the history of optics and the scientific heritage from the seventeenth and eighteenth centuries.

IVANO DAL PRETE received his PhD in History of European Society in 2005 from the University of Verona. He has held teaching positions in the history of science / history of medicine programs at Yale University and at the University of Minnesota, and he is presently Associate Research Scholar at the Italian Academy, Columbia University. He is the author of *Scienza e Società nel Settecento Veneto: Il Caso Veronese 1680–1796* (Milano, 2008) and is currently working on a book project on earth history in the early-modern period.

Inga Elmqvist Söderlund, PhD, has been Curator and Director of the Observatory Museum in Stockholm since 1996. She has written and curated several exhibitions on early-modern art and science. She is presently working on the project "Scientific Instruments in Eighteenth-Century Europe – Display, Visual Effect and Aesthetic Experience?" based at the Museum of the History of Science in the University of Oxford.

David Felismino is a PhD student in cultural modern history at the Faculdade de Ciências Sociais e Humanas of the Universidade Nova de Lisbonne. He is a research member of the Centro de História de Além Mar, and he is associated to the European network "La Corte en Europa". His research interests include several themes in cultural history, especially court culture and history of everyday life, consumption and networks of cultural exchange, mainly between the sixteenth and nineteenth centuries. He is a researcher on the project "On the Instruments' Trail: Exploring Royal Cabinets in Portugal".

Hans Hooijmaijers is Head of Collections and Vice-Director at Museum Boerhaave in Leiden. He joined the Museum after studying physics and his main interests are Christiaan Huygens, the eighteenth-century Leiden Physics Cabinet and the history of the Leiden Observatory. His latest writing projects were on Dutch orreries, the nineteenth-century Leiden Observatory and Dutch navigation.

Marta C. Lourenço has been a researcher at the Museum of the University of Lisbon since 1998. Her research interests include university collections, the history of collections and scientific heritage. She is a research member of the Interuniversity Research Centre of the History of Science and Technology and associate researcher of the Centre d'Histoire des Techniques et de l'Environnement, Paris. She coordinates the research project "On the Instruments' Trail: Exploring Royal Cabinets in Portugal". She is Vice-President of "Universeum" (European Network for University Heritage) and a board member of ICOM-Portugal.

Ad Maas is a curator at the Museum Boerhaave. He specializes in the history of Dutch physics between 1850 and 1945, Albert Einstein, Dutch scientific culture between 1750 and 1900, the Museum Boerhaave and scientists in World War II. He published on 's Gravesande in the volume co-edited with Eric Jorimk, Newton and the Netherlands: how Newton was fashioned in the Dutch Republic (Leiden, 2012).

SOFIA TALAS is Curator of the Museum of the History of Physics at the University of Padua. Her main research interests are in the history of scientific instruments and the history of physics. She is President of "Universeum" (European Network for University Heritage) and a board member of the History of Physics Group of the European Physical Society.

MARTIN WEISS graduated from the University of Utrecht in history and philosophy of science in 2008, after previously studying physics at RWTH Aachen University. He is currently working on a dissertation on the history of the Teyler Museum in the nineteenth century at the University of Leiden.

EWA WYKA is Curator of the Department of History of Science and Scientific Instruments at the Jagiellonian University Museum in Kraków. Her main subjects of interest are eighteenth-century physics and chemistry instruments, eighteenth-century collections in Poland, royal collections and history of science in Poland.

HUIB J. ZUIDERVAART is a historian of science at the Huygens Institute for the History of the Netherlands of the Royal Netherlands Academy of Arts and Sciences in The Hague. He specializes in the history of scientific instruments and of astronomy in the seventeenth- and eighteenth-century Netherlands. For his publications, see www.huygens.knaw.nl/en/zuidervaart/.

INTRODUCTION

Experimental science is generally thought to be a creation of the seventeenth century, when it started to be practised within a growing number of scientific academies. Experiments and demonstrations about various aspects of the natural world were regularly held, attended by the academies' members. It was only in the last years of the century, that lessons in natural philosophy, illustrated by experiments, started to be introduced in a very few universities, within philosophy courses. "Experimental natural philosophy" included hydrostatics, pneumatics, mechanics, optics, acoustics, magnetism and chemistry, as well as the study of minerals, animals or plants.

At the beginning of the eighteenth century, John Keill and William Whiston, both pupils of Newton, introduced new experimental lectures in Cambridge and Oxford and, under Newton's influence, experiment-based lectures were offered in London to a wide public. Most of the lecturers – among the most prominent of these, we might mention John Theophilus Desaguliers – were or became Fellows of the Royal Society and the lectures drew widely on what had been performed during previous meetings of the Society. However they presented only experiments on mechanics, hydrostatics, pneumatics and optics: the field of "experimental philosophy" was becoming more circumscribed. The resources provided by the scientific instrument market, which was flowering in London at that time, were crucial for the success of the new lectures.

The impact of such lectures was rapid and, after their introduction in England, they were brought to the Netherlands by Willem 's Gravesande and Pieter van Musschenbroek, both strenuous supporters of Newtonianism. From England and Holland, the new experimental philosophy practice spread and within a few decades reached the whole of Europe and the British colonies in North America. Chairs in the subject were created at universities, thus marking the entry of experimental science into university teaching. Not only was a new scientific practice being set up, but physics in its current sense was to emerge as a specific field of knowledge, opening also the way to the emergence of other disciplines (chemistry, etc.) in their current meaning.

In relation with the new way of lecturing and demonstrating, treatises of experimental physics were published, more and more instruments were specifically designed, and hundreds of collections of scientific instruments – the so called "cabinets of physics" – were created throughout Europe in a variety of contexts. Physics acquired in those years an extraordinary popularity, not only among specialists, but in the "salons" and in royal courts as well. Science was turning from private to public.

The growing interest among historians of science in the development of a public practice of experimental philosophy in the eighteenth century, based around lectures, demonstrations and cabinets of physics, is an important opportunity for instrument historians and curators. The attraction of the topic for general science historians is not difficult to understand, since it engages with many of the current interests of science historiography. Material culture is at its core; performance, experiment and demonstration are characteristic of its "offer" to its public; commerce is its *raison d'être*, supported by religious sentiment, education and sociability; its spatial accommodation, be it a lecture room, a museum cabinet or a shop, gives shape to the modes of learning and knowing it conveys; its variety of institutional settings, be they courtly, academic or learned, differentiate what is, in theory, a collective engagement with the natural world. Neither is an institutional location essential: the dynamic relationships between institutional, commercial and domestic settings are a further, compelling facet of the emergence of this experimental culture, that set the basis for further crucial developments of physics in the nineteenth century.

The geographical spread of cabinets of experimental philosophy throughout Europe in the eighteenth century offers an opportunity to study the growing success of a scientific canon – an orthodoxy of method and social organisation, as well as of scientific content. How did it come about that so many collections of instruments sprang up across Europe (and in America, though this set of essays is restricted to Europe)? Why was there so much imitation and consensus? What roles were played by publication and the acceptance of written authorities, by instrument making and its international trade, by travel and correspondence?

On the other hand, how was the consensus of practice accommodated to local interests and resources? How was it modified with distance and time? Should we draw a geography of centres and peripheries, lines of influence, or contours of activity and engagement? How might any such map reflect patterns of power, trade, commerce and learning?

This volume of essays, the third in the series "Scientific Instruments and Collections", published under the auspices of the Scientific Instrument Commission and the general editorship of Giorgio Strano, aims to contribute to this topic by illustrating something of the variety of eighteenth-century cabinets and something of their fate in subsequent epochs – whether that was to be accommodation and survival or irrelevance and extinction. We hope that this volume will form a significant part of a larger-scale project, covering all European countries, together with the United States and wherever else early cabinets were to be found, aimed at a much more complete understanding of the topic.

The papers derive, in the main, from contributions to a special session organised by the volume editors within the programme of the 30th Scientific Instrument Symposium, held in the Astronomisch-Physikalisches Kabinet in Kassel, Germany, in September 2011. We are grateful to Karsten Gaulke and Bjoern Schirmeier, as well as the other members of the organising committee, for a generous allocation of time to this session. We are grateful also to the contributors, whose papers, we believe, move important aspects of the subject forward in ways that merit publication.

Taking his point of departure from the auction catalogue of the sale in 1889 of the collection once under the care of the society *Felix Meritis* in Amsterdam, Huib Zuidervaart offers an impressively broad account of cabinets in the Netherlands. Striking for their ubiquity in the period and for the variety of their collectors and owners, it is remarkable also that the traces they have left are so contingent on the vagaries of circumstance – only a very few survive in any material form and, for a survey such as Zuidervaart's, the historian must be resourceful in chasing documentary records. The number of collections covered here allows Zuidervaart some valuable and well-grounded generalisations, at least in respect of the Netherlands. It allows him also to address the question of whether and to what extent such cabinets were precursors of physics laboratories, granted that the fate of many was to be expelled by later physicists.

We stay in the Netherlands for an account by Hans Hooijmaijers and Ad Maas of one of the earliest cabinets, the Leiden Cabinet of Physics, founded in 1675. An example of a teaching cabinet in a University setting, its prodigious reputation was principally owing to an outstanding director, Willem Jacob 's Gravesande. While the influence of his presentation of the British or, more specifically, Newtonian approach to experimental philosophy had an international reach, we learn of the importance of local needs within the University and the personal ambitions of early directors in shaping the character and contents of this famous cabinet.

Padua and Oxford provide examples of how different university settings are themselves very distinctive contexts for building teaching collections. Sofia Talas shows that in Padua, the initiative and continuing support of a centralised agenda for reform, originating in the government of the Venetian Republic, was the impetus for developing an extensive collection.

By contrast, the teaching collection in Oxford – of a comparable size, but now totally disappeared – was created by the personal initiatives of successive professors and its demise, as we learn from Jim Bennett's account, followed quickly after the intervention of the University authorities. Beyond some differences, similarities in the way of teaching in both places are striking, both in the structure of the lessons and in the attention paid by the lecturers to content and presentation.

Staying within university context, Tiemen Cocquyt takes us back to Leiden, for a longer and more inclusive view of the activity involving the cabinet there and the concerns of the professors in charge of it. He shows that these could extend beyond an academic interest in experimental natural philosophy, into entrepreneurial and commercial projects, where the professors were not so securely grounded. Illustrated by models of machines included in the cabinets, the social involvement of experimental philosophy practitioners and "enthusiasts", be they scholars, teachers or kings, is a recurring theme in several other papers in the volume.

Attention then moves beyond an academic setting, to deal with collections operating under other forms of patronage. Inga Elmqvist Söderlund begins with a cabinet funded through a private endowment and housed in the Riddarhuset in Stockholm, the house of the Nobility, one of the four governing estates of Sweden. Though outside the University, this was essentially an educational initiative, later taken over by the Royal Swedish Academy of Sciences.

If Riddarhuset had aspects of the private and the public, while operating mainly in a public setting, Paola Bertucci describes a mirror image – a courtly cabinet, established by Ferdinando Spinelli, prince of Tarsia, that was open to the public. As well as a description of the museum and library, we have here an analysis of the social roles played by this cabinet and of the tensions created by the conflicting ambitions of those involved – the patron Spinelli and members of the academy he created to make use of his cabinet.

The cabinet described by Marta Lourenço and David Felismino is unequivocal in its courtly identity. It was created in the 1760s in the Palace of Ajuda in Lisbon for the education of the two sons of Queen Maria I, Princes José and João. Despite this limited purpose, it shared many of the characteristics of cabinets in universities and academies throughout Europe, so was a manifestation of a broad experimental culture.

Ivano Dal Prete's account of another courtly cabinet, that of the Gazola family in Verona in the late 1700s, introduces a further public use of a private collection. As well as being open to visitors, we learn that the collection could be a tool for forwarding the cultural and political agendas of its owners, and again we see the relevance of cabinets to broader interests in the historiography of science.

Ewa Wyka extends this pattern of experimental culture eastwards with a survey of cabinets in Poland, covering aristocratic collections as well as those in universities and colleges run by religious orders. As with Portugal in the west, it is clear that here are further instances of a European movement, adapted, as it is everywhere, whether "central" or "peripheral", to local contingencies.

With cabinets of experimental philosophy multiplied throughout Europe, it might be wondered why so few seem to survive. One account of a survival explains how a collection has come down to us today. Martin Weiss tells the story of the Teyler Museum in Haarlem, which has become a fully public museum in partnership with a fine art collection. The cabinet is extraordinarily rich and extensive, but if we wonder why there are so few such survivors, we have only to read the precarious twists and turns in this detailed narrative.

Also going beyond the eighteenth century, Paolo Brenni's paper clearly shows how the Cabinet of Physics set up in Florence in 1775 turned out to be insufficient for the needs of scientific and technical education. It then focuses on the new, purposely-created collection of the Istituto Tecnico Toscano, thus providing an outline of some of the features of the cabinets of physics that were set up in the nineteenth century.

The range of these essays illustrates how an institution – the cabinet of physics – could maintain a significant degree of intellectual identity and a large measure of shared content, while arising and surviving (at least for a time) in very different contexts and meeting different needs and ambitions. While this is not unlike other institutions – the orchestra, for example, could be a parallel case, with similar characteristics – we are more used to thinking of science spreading by conquest or conversion, with little appetite for local accommodation. What we see from these studies is that the social resources and values of the cabinet were adaptable, while its material resources were sufficiently prominent to support a shared identity and intellectual culture. Common textbooks were an important part of this material resource; just as important were common instruments.

Jim Bennett
Sofia Talas

CABINETS FOR EXPERIMENTAL PHILOSOPHY IN THE NETHERLANDS

Huib J. Zuidervaart

Introduction

On 2 December 1889 in Amsterdam a remarkable auction was held. A colossal collection of scientific instruments and related tools was sold, the origin of which – at least in the auction catalogue – was left in the dark.[1] Only by comparing the content of this catalogue with the latest inventory of the former cabinet of experimental philosophy of the Amsterdam Society *Felix Meritis* (liquidated shortly before, in June 1889) can it be established that a good part of the Society's renowned collection of scientific instruments came under the hammer.[2] This in itself is remarkable, since at its liquidation the board of *Felix Meritis* had expressly stipulated that only those goods of the Society would be donated to municipal bodies, of which it was evident that these institutions really would benefit by these gifts. The library of Amsterdam University, for instance, regarded the voluminous stock of books donated by *Felix Meritis* as a very valuable addition to its collection; likewise, the *Felix Meritis* plaster casts were accepted with great enthusiasm by the Rijksacademie voor Beeldende Kunsten (National Academy of Visual Arts) (*Figs. 1 and 2*), and in this same manner it was agreed with the director of the Physics Laboratory of the Amsterdam Municipal University that the scientific instruments of the *Felix Meritis* Society would be included in this Laboratory as a complete and indivisible collection.[3] However, the relevant professor of physics, Johannes Didericus van der Waals, being a genuine theorist, was not

[1] *Catalogus van een kolossale kollectie instrumenten en gereedschappen, waarvan de publieke verkoping zal plaatsvinden op maandag 2 december 1889*, Amsterdam, 1889; City Archive Amsterdam, VC-21. See also Huib J. Zuidervaart, Rob H. van Gent, *Between Rhetoric en Reality. Astronomical Practices at the Observatory of the Amsterdam Society 'Felix Meritis' 1786–1889*, Hilversum, 2013.

[2] "Catalogus van het Physisch Kabinet der Maatschappij Felix Meritis", 1871; Amsterdam University, Special collections Department. In 1889 probably also some obsolete instruments from the cabinet of the former Amsterdam *Athenaeum Illustre* were auctioned. Cf. *Van Athenaeum tot Universiteit. Geschiedenis van het Athenaeum Illustre in de negentiende eeuw*, Amsterdam, 1927, pp. 63–64.

[3] Cf. *Gemeenteblad Amsterdam, tweede afdeling: verslag der vergaderingen van den gemeenteraad, enz. 1889*, Amsterdam, 1889, pp. 369–370 [meeting 12 June 1889].

Fig. 1. The building of the *Felix Meritis* Society at the Amsterdam Keizersgracht, c. 1800.

keen on experimental equipment. Prior to the transfer of the instruments he had written to the board of *Felix Meritis* that, in his eyes, these instruments only had some historical value.[4] So, it seems plausible to suppose that Van der Waals personally has given impetus to the instrument sale, mentioned before.[5] Only a small number of notable instruments, such as three planetariums, remained at the University.[6]

[4] Amsterdam, City Archive, Archive *Felix Meritis*, inv. nr. 59.

[5] According to tradition in 1889 a large part of the *Felix Meritis* instruments was bought by Nicolaas Groenendijk, a former *Felix Meritis* member. His son and heir N.C.A.J. Groenendijk organised in 1958 an auction of more than 500 scientific instruments. Some 120 instruments were obtained by the forerunner of Museum Boerhaave in Leiden. Around 140 instruments were bought by the American dealer Alvin C. Detwiler. These instruments were purchased *en bloc* by the Burndy Library of the Dibner Institute in Norwalk (Connecticut). In 2006 this institute was transferred to the Huntington Library in San Marino (California). Cf. Peter de Clercq, "De familie Groenendijk", in de Clercq, *Het koperen kabinet. Schatkamers van de wetenschap, 1550–1950*, Leiden, 1994, pp. 54–56, and an internal (undated) document "The Groenendijk Collection", sent to the present author by David McGee from the Burndy Library. See also: Willem Hackmann, "In the Hall of Physics", *Bulletin of the Scientific Instrument Society* 67 (2000), pp. 1–2, who claims that some Groenendijk-objects were bought by the Smithsonian Institution, National Museum of American History, Washington D.C.

[6] In the early twentieth-century some remaining *Felix Meritis* instruments found their way to the Nederlands Scheepvaartmuseum in Amsterdam and the Rijksmuseum voor de Geschiedenis der Natuurwetenschappen in Leiden (today renamed Museum Boerhaave).

Fig. 2. Plaster, depicting an arrangement of scientific instruments in the staircase of the *Felix Meritis* building in Amsterdam, designed in 1789 by the architect Jacob Otten Husley (who shortly before had made a similar plaster in Teyler Museum in Haarlem). Presented from the left to the right are: (parts of) the hose and nozzle of a fire extinguisher, a plate electrical machine, a Nuremberg-type (?) microscope, an Archimedean screw, a globe, a folio book, a reflecting telescope, the ocular side of a refracting telescope, an anamorphic mirror or jar, a stand for a lens or mirror, a two piston air pump, and in the front two leather clamps, probably for the fastening of the hose of the fire extinguisher (Photo by the author, 2006).

The last inventory of the *Felix Meritis* instrument cabinet was drawn up in 1871. This list enumerated a total of 463 scientific instruments.[7] Compared with the 280 scientific instruments then preserved in the – today far better known – Teyler Museum in Haarlem, the *Felix Meritis* collection of instruments was not only numerically larger, but also of equal quality.[8]

---

[7] "Catalogus van het Physisch Kabinet...", op. cit. (n. 2).

[8] "Catalogus der natuurkundige werktuigen in het Museum van Teylers Stichting", made up in July 1859, printed in: Gerard L'Estrange Turner, *The Practice of Science in the*

Why did these general learned societies posses cabinets with scientific instruments? Why did these institutions ever decide to bring together and use such collections? There were more: in the eighteenth and early nineteenth centuries, one stumbled on societies and individuals with cabinets of experimental philosophy. Why had this phenomenon occurred and how did these instruments function over time? And finally, why are these instrument collections only rarely preserved, in contrast to collections with pieces of art or other historical artefacts? What made the fate of these – often very nicely crafted – instruments so different? For what today still can be seen in museums, is largely determined by mere chance. Actually, only in the current Teyler Museum there is a continuum with the original eighteenth-century collectors. Other Dutch museums with scientific instruments have a much more recent – twentieth-century – genesis. In short, what can we tell about the origin, development and fate of the scientific instrument collections in the Netherlands?

The Origin of the Cabinet of Experimental Philosophy

The specialised cabinet with solely scientific instruments is typically a product of the eighteenth century. Certainly, the *Kunst und Wunderkammer* of the late Renaissance also devoted special attention to scientific (mostly mathematical) instruments, but those were usually tools that could perform specific tasks. With these devices something could be measured, such as lengths, angles or areas. The legitimacy of these objects traditionally was derived from this practical applicability. But in the seventeenth century instruments emerged with a previously unseen meaning and function: contemporaries often referred to them as "philosophical instruments". These devices did not measure anything, but they made things visible, or clarified some principles. Initially it was unclear what exactly instruments such as the telescope, microscope, thermoscope, baroscope, hygroscope, air pump and – somewhat later – the electrostatic generator added to scientific knowledge. In the course of time some evolved into real measuring devices, for instance the barometer and thermometer. But as tools for natural philosophy, these new instruments initially had

---

Nineteenth Century. Teaching and Research Apparatus in the Teyler Museum, Haarlem, 1996, pp. 319–331.

to overcome significant resistance. How should their results be used and interpreted? What gave them legitimacy?[9]

Studies by, for example Shapin and Schaffer, have shown that these instruments have won their place in the scientific enterprise as a result of a complex process involving several social and local parameters. In the construction of scientific facts, all kinds of personal, instrumental and socio-cultural interactions play a role. Authors like Golinsky, Stewart, Sutton, Walters and others, have discussed various aspects of this cultural phenomenon.[10] Within a rather short period natural philosophy evolved from a relatively obscure activity of a handful of scholars, to a broad-based activity of many people. In the early eighteenth century "experimental philosophy" became a valuable activity in which many wanted to participate.[11]

The Cabinet of Experimental Philosophy in the Dutch Republic

The Netherlands clearly participated in this general European development.[12] In the 1730s a wave of popularization of "Newtonian" philosophy went through the Dutch Republic. Where previously collectors had focused on bringing together natural history specimens and

---

[9] Cf. Albert van Helden, Thomas L. Hankins, "Instruments in the History of Science", *Osiris* 9 (1994), pp. 1–6; Deborah J. Warner, "What is a scientific instrument, when did it become one, and why?", *British Journal for the History of Science* 23 (1990), pp. 83–93.

[10] Cf. Jan Golinsky, *Science as Public Culture. Chemistry and the Enlightenment in Britain, 1760–1820*, Cambridge 1992; Simon Schaffer, "Natural Philosophy and public spectacle in the 18th century", *History of Science* 21 (1983), pp. 1–43; Steven Shapin, "A Scholar and a Gentleman: The Problematic Identity of the Scientific Practitioner in Early Modern England", *History of Science* 29 (1991), pp. 279–327; Larry Stewart, *The Rise of Public Science. Rhetoric, Technology, and Natural Philosophy in Newtonian Britain, 1660–1750*, Cambridge, 1992; Geoffrey V. Sutton, *Science for a Polite Society. Gender, Culture & the Demonstration of Enlightenment*, Colorado, 1995; Alice N. Walters, "Conversation pieces: Science and Politeness in Eighteenth-century England", *History of Science* 35 (1997), pp. 121–154.

[11] Cf. G.L'E. Turner, "The Cabinet of Experimental Philosophy", in Oliver Impey, Arthur MacGregor (eds.), *The Origins of Museums. The Cabinet of Curiosities in Sixteenth- and Seventeenth-century Europe*, Oxford, 1985, pp. 214–222.

[12] Cf. Harold J. Cook, *Matters of Exchange. Commerce, Medicine, and Science in the Dutch Golden Age*, New Haven-London, 2007, and Eric Jorink, *Reading the Book of Nature in the Dutch Golden Age, 1575–1715*, Leiden-Boston, 2010. For the Southern Netherlands (nowadays Belgium), see: Claude Sorgeloos, "Les cabinets d'histoire naturelle et de physique dans les Pays-Bas autrichiens et à Liège", in Marie-Thérèse Isaac, Claude Sorgeloos (eds.), *La diffusion du savoir scientifique XVIe–XIXe siècles*, Bruxelles, 1996, pp. 125–230.

artificialia,[13] now a new form of collecting emerged: the "cabinet of experimental philosophy". Wealthy merchants, bankers and other enthusiasts, especially in the urbanized part of the country, started to build such an instrument collection, often as an extension of their library.

On the causes of this popularity much has been written.[14] One reason was simply the success that the experimental method had managed to achieve since its genesis in the seventeenth century. The discoveries these instruments had brought to light appealed to the imagination of the public. It also made people aware of the manipulability of nature. In the Dutch Republic another reason for the vogue of experimental physics was the acceptance, from 1715 onwards, of "Newtonianism" as a guiding philosophical principle in nature. In 1687 Newton had published his *Philosophiae Naturalis Principia Mathematica*. Then, the book was noted only by a small circle of highbrow scholars. But the reprints of 1713 (in Cambridge) and 1714 (in Amsterdam) increased the spread of this important text enormously. Crucial also was the addition to this second edition of the so-called "General Scholium". In this text Newton left room for a possible divine intervention in nature. This theological component was probably the most important motivation for the acceptance of Newtonianism in the Netherlands.[15] Although Newton searched for the laws of nature, for him those laws were laid into nature only by the grace of God, as a token of His reliability. This view was in strong contrast to the atheistic rationalism, preached by some scholars at the turn of the seventeenth century. Newton's epigone William Derham even introduced a name for this

---

[13] Ellinoor Bergvelt, Debora J. Meijers, Mieke Rijnders (eds.), *Kabinetten, galerijen en musea. Het verzamelen en presenteren van naturalia en kunst van 1500 tot heden*, Zwolle, 2005, pp. 15–68 and 101–128. See also: Bert van de Roemer, "Neat Nature: The Relation between Nature and Art in a Dutch Cabinet of Curiosities from the Early Eighteenth Century", *History of Science* 42 (2004), pp. 47–84.

[14] Harry A.M. Snelders, "Professors, amateurs and learned societies: the organisation of the natural sciences", in Margaret C. Jacob, Wijnand W. Mijnhardt (eds.), *The Dutch republic in the 18th century: Decline, enlightenment, and revolution*, Ithaca, 1992, pp. 308–323; Lissa Roberts, "Going Dutch: Situating Science in the Dutch Enlightenment", in W. Clark, J. Golinsky, S. Schaffer (eds.), *The Sciences in Enlightened Europe*, Chicago-London, 1999, pp. 350–388; H.J. Zuidervaart, "Science for the public: The translation of popular texts on Experimental Philosophy into the Dutch language in Mid-Eighteenth Century", in Stefanie Stockhorst (ed.), *Cultural Transfer through Translation. The Circulation of Enlightened Thought in Europe by Means of Translation*, Amsterdam-New York, 2010, pp. 231–262.

[15] E. Jorink, H.J. Zuidervaart, "The Miracle of Our Time. How Isaac Newton was fashioned in the Netherlands", in E. Jorink, A. Maas (eds.), *Newton and the Netherlands. How Isaac Newton was fashioned in the Dutch Republic*, Leiden, 2012, pp. 13–65. See also: Rienk Vermij, "The Formation of the Newtonian Philosophy: The Case of the Amsterdam Mathematical Amateurs", *British Journal for the History of Science* 36 (2003), pp. 183–200.

religious "Newtonian" method of considering nature: "Physico-Theology" (a term he used for a book, published in 1713). As Newtonians like Derham and his Dutch counterpart Bernard Nieuwentyt particularly praised God as the ultimate and omnipotent Creator of the universe, the conservative Dutch clergy no longer had an argument against the study of nature. As a result, the physico-theological approach contributed significantly to the popularization of experimental physics in the Netherlands. After all, philosophical instruments provided a broader perspective on natural phenomena and thus offered the possibility of pious contemplation of the greatness of God's creation.[16]

The Leiden professor Willem Jacob 's Gravesande was one of the most prominent advocates of Newtonian physics. His Handbook *Physices elementa mathematica, experimentis confirmata sive introductio ad philosophiam Newtonianam* (Mathematical principles of physics, supported by experiments, presenting an introduction into the Newtonian Philosophy), published in Latin in 1719–1721, became very successful throughout Europe and attracted many students to Leiden University. The demonstration instruments that 's Gravesande developed for his book, in cooperation with the Leiden instrument maker Jan van Musschenbroek, were produced in vast numbers. They found their way to numerous instrument cabinets in eighteenth-century Europe.[17] Finally, there were also more mundane reasons for the popularization of physics, such as entertainment and social status. Very successful in that respect was the English Newtonian John Theophilus Desaguliers, who in the years 1729–1732 held a series of public lectures on physics and astronomy in various Dutch cities.[18] According to the Leiden professor Pieter van Musschenbroek the lessons of the "far-famed philosopher Desaguliers" had cultivated a

---

16 See for instance the physico-theological poem, refering to the reflecting telescope, in: Catherina Pietersdochter Braconier-De Wilde, *Bespiegelingen over Gods Kerk- en Waereldbestier* [...] *Nevens Zielsverlustiging in het beschouwen van Aarde, Lucht en Sterrenhemel*, Amsterdam, 1751, p. 162. See for Dutch physico-theology in general: Jan Bots, *Tussen Descartes en Darwin: geloof en natuurwetenschap in de achttiende eeuw in Nederland*, Assen, 1972.

17 P. de Clercq, *The Leiden Cabinet of Physics. A Descriptive Catalogue*, Leiden, 1997; de Clercq, *At the Sign of the Oriental Lamp. The Musschenbroek Workshop in Leiden, 1660–1750*, Rotterdam, 1997. For the later period, see: de Clercq, "In de schaduw van 's Gravesande. Het Leids Physisch Kabinet in de tweede helft van de 18e eeuw", *Tijdschrift voor de Geschiedenis der Geneeskunde, Natuurwetenschappen, Wiskunde en Techniek* 10 (1987), pp. 149–173.

18 Marius J. van Lieburg, "De Geneeskunde en natuurwetenschappen binnen de Rotterdamse geleerde genootschappen uit de 18e eeuw", *Tijdschrift voor de Geschiedenis der Geneeskunde, Natuurwetenschappen, Wiskunde en Techniek* 1 (1978), pp. 14–22, 124–143.

large number of enthusiasts.[19] In cities like Amsterdam, Haarlem, Dordrecht, Schiedam and Middelburg, Desaguliers' activity even resulted in the founding of a few informal scientific societies, two of which even had each their own building, instrument cabinet and observatory.[20] Desaguliers' fans soon became customers: at first of Desaguliers himself, who brokered the sale of many scientific instruments, but also of a growing number of English and Dutch instrument makers.[21]

Collectors: Nature and Distribution

The products of these instrument makers were brought together in a rising number of instrument collections. How extensive this tendency was, cannot be said with certainty. It definitely was not a marginal phenomenon. So far, I have collected information on some five hundred Dutch collectors of scientific instruments. Graph 1 (*Fig. 3*) representing the known number of auction sales and inventories containing scientific instruments, shows that this number peaks in the early nineteenth century. After about 1870 almost all private instrument cabinets disappear from the scene. Before the turn of the seventeenth century mathematical or astronomical instruments were almost the only ones to be mentioned. The genuine "philosophical" instruments emerge mostly in the second quarter of the eighteenth century. Devices for static electricity entered the scene only in the second half of the century.[22]

When we select cabinets with at least 25 instruments, then (in the present state of the known material) "only" some 70 instrument collections remain.[23] However, this gives only an indication of the size of the

[19] Pieter van Musschenbroek, *Beginsels der Natuurkunde, Beschreeven ten dienste der Landgenooten*, Leiden, 1739, preface.

[20] Bert C. Sliggers, "Honderd jaar natuurkundige amateurs in Haarlem", in Anton Wiechmann, Lodewijk C. Palm (eds.), *Een elektriserend geleerde: Martinus van Marum, 1750–1837*, Haarlem, 1987, pp. 67–102; H.J. Zuidervaart, *Van 'Konstgenoten' en Hemelse Fenomenen. Nederlandse Sterrenkunde in de Achttiende Eeuw*, Rotterdam, 1999.

[21] Cf. Maria Rooseboom, *Bijdrage tot de Geschiedenis der Instrumentmakerskunst in de noordelijke Nederlanden tot omstreeks 1840*, Leiden, 1950. See also: P. de Clercq, "The instruments of Science: the Market and the Makers", in Klaas van Berkel, Albert van Helden, Lodewijk C. Palm (eds.), *A History of Science in The Netherlands: Survey, Themes, and Reference*, Leiden, 1999, pp. 311–331.

[22] L. Roberts, "Science becomes Electric. Dutch interaction with the Electrical Machine during the Eighteenth Century", *Isis* 90 (1999), pp. 680–714.

[23] Only a limited number of Dutch cabinets for experimental philosophy have been discussed. See for example: J.A. Bierens de Haan, "Het huis van een 18e eeuwse 'Mercator

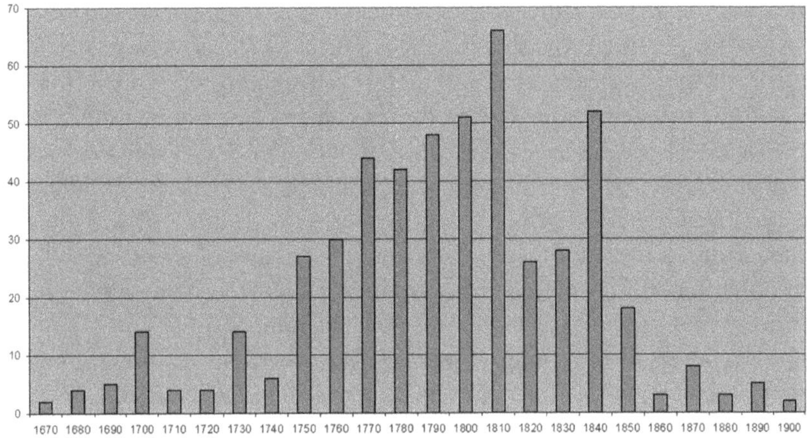

Fig. 3. Graph 1: Number of auction sales and inventories containing scientific instruments, clustered in intervals of ten years during the period 1670–1900.

phenomenon. With regard to ephemeral publications, such as auction catalogues, one always has to remember that mainly chance determines what has been preserved throughout time. For instance it is estimated for Dutch book-sale catalogues that only between 5% to 10% of these catalogues have survived. Many private instrument collections were sold together with the libraries of their owners. Specialized auction catalogues were printed only when large instrument collections were sold. To my knowledge, the first specialized sale of a Dutch cabinet of experimental philosophy took place in 1724, when the cabinet of the Leiden professor Wolferd Senguerd came under the hammer.[24] Together with his colleague Burchard de Volder, Senguerd had been the founder of the *Theatrum*

Sapiëns' ", *Jaarboek van het genootschap Amstelodamum* 49 (1957), pp. 110–128 on the collection of Anthony Bierens, auctioned in 1747; Paul Hoftijzer, "An eighteenth-century amateur of books and science in Warmond", in Jan Frans van Dijkhuizen et al. (eds.), *Living in posterity. Essays in honour of Bart Westerweel*, Hilversum, 2004, pp. 147–156 on the collection of Abraham Edens, auctioned in 1765; H.J. Zuidervaart, "Mr. Johan Adriaen van de Perre (1738–1790). Portret van een Zeeuws regent, mecenas en liefhebber van nuttige wetenschappen", *Archief. Mededelingen van het Koninklijk Zeeuws Genootschap der wetenschappen* (1983), pp. 1–169 on the collection auctioned in 1798, and Zuidervaart, "Egbert Philip van Visvliet (1736–1799). Een Middelburgs verzamelaar van vrienden, naturalia en fysische instrumenten", *Zeeland. Tijdschrift van het Koninklijk Zeeuws Genootschap der wetenschappen* 3 (1994), pp. 87–95 on the collection auctioned in 1799.

24 Unfortunately no copy of the catalogue of this *Musaeum Instrumentorum Mechanico-Physico Mathematicorem* has been found. See: *Leydse Courant* 5, 10 and 17 May 1724. Earlier an attempt had been made to sell the collection as a whole. See. *Leydse Courant* 11, 16 and 18 February 1724.

Physicum of Leiden University.[25] Therefore, it is no surprise that he possessed an "excellent" instrument cabinet. In those days a natural philosophy professor was expected to be largely self-supporting in respect to the scientific instruments he needed for his courses. It was only in 1742, after the death of professor 's Gravesande, that Leiden University decided to buy his cabinet, so that the continuity of his famous lectures – highlighted by practical demonstrations – was guaranteed. In this way for the first time a Dutch institution founded a genuine cabinet of experimental philosophy. Before that date Leiden University possessed only a handful of instruments, among which were several air pumps.[26]

An analysis of the group of people who owned an instrument collection, based on 199 identified owners from the period 1670–1860, shows that approximately one-fifth used their collection for educational purposes (*see Table 1*). The majority of the cabinet owners may be characterized as "enthusiasts", although in the period of study it still is rather difficult to distinguish between "professionals" and "enthusiasts" (*Fig. 4*). The oldest known specialized auction catalogue of the instrument cabinet of a real "enthusiast" dates from 1743, describing the instrument collection of Nicolaas van der Haer, registrar of the Supreme Court of Holland in The Hague.[27] Looking at the regional distribution of the known instrument cabinets, it is clear that the phenomenon occurred especially in urbanized areas of the country (*see Table 2*).

Remarkably, the list of cities in *Table 2* is headed by non-university residences like Amsterdam and The Hague. Interestingly, these two places accommodated almost half of the known cabinets of experimental philosophy. That the wealthy merchant milieu of Amsterdam is high on the list, can hardly be seen as a surprise, but the high score of the foremost politically-oriented city The Hague (the seat of the States-General) needs

---

[25] Gerhard Wiesenfeld, *Leerer Raum in Minervas Haus, Experimentelle Naturlehre an der Universität Leiden, 1675–1715*, Amsterdam, 2002.

[26] De Clercq gives the following dates as the prime moments of acquisition of a scientific instrument (mostly an air pump) by Dutch universities: Leiden (1674); Groningen (1695); Franeker (c. 1701) and Utrecht (1705). Cf. De Clercq, *Sign of the Oriental Lamp...*, op. cit. (n. 17), pp. 134–149. Until 1804 Harderwijk University only possessed a large burning mirror. Cf. Cornelus Philippus Jacobus Penning, "Gedenksteen te Harderwijk en bouw eener kamer voor de bewaring van physische instrumenten", *Bijdragen en mededeelingen der Vereeniging 'Gelre'* 33 (1930), pp. 169–172.

[27] *Catalogus van een keurige partye physische en mathematische instrumenten en andere fraaje rarititeiten, nagelaten door wylen den Wel. Ed. Gestr. Heer en Meester Nicolaas van der Haer*, 's Gravenhage, 1743 (copy kept in Museum Meermanno Westreenianum, The Hague).

Table 1. Dutch collectors of scientific instruments (1670–1860).

Categories	Professions	Number	%
Education	Professor in philosophy	19	
	Lector in philosophy	16	
	School master	1	
		36	18
Practical occupations	Instrument maker	12	
	Mathematician	11	
	Military	6	
		29	15
Medicine	M.D.	25	
	Professor in Medicine	11	
	Surgeon	2	
	Apothecary	1	
		39	20
Remaining	Magistrate	36	
	Merchant	19	
	Lawyer	16	
	Civil servant	8	
	Preacher	6	
	Diplomat	4	
	Book trader	4	
	Professor in Law	2	
		95	48
		199	**100**

Table 2. Regional distribution of the collectors in Table 1.

Place	Number	%
Amsterdam	47	24
The Hague	39	20
Leiden	27	14
Leeuwarden	14	7
Middelburg	12	6
Haarlem	8	4
Utrecht	7	4
Delft	4	2
Dordrecht	4	2
Franeker	3	2
Other	17	9
Unknown	17	9
	199	**100**

Fig. 4. Demonstration of an electrical plate machine in the Physics Theatre of *Felix Meritis*, probably by the lector Henricus Aeneae (Drawing by P. Barbiers and J. Kuyper, 1794, engraved by Reinier Vinkeles, 1801).

more interpretation. This result probably must be seen in the light of the establishment, in 1747, of a semi-royal court in The Hague. After a long period without a *stadholder* (since 1702), this semi-monarchal position had been restored in the entire Dutch Republic in 1747. Until that year only two Northern provinces (Friesland and Groningen) had maintained this function of (among others) commander-in-chief of the armed forces. But after the restoration of 1747 the court of *stadholder* William IV of Orange-Nassau moved from Leeuwarden to The Hague. Since the 1730s William had collected an impressive cabinet of experimental philosophy, with state-of-the-art instruments made on both sides of the North Sea.[28] As son-in-law of the English (and Hanoverian) king George II, William IV had frequently visited London. There he had followed lessons in natural

[28] P. de Clercq, "Science at Court: the Eighteenth-century Cabinet of Scientific Instruments and Models of the Dutch Stadholders", *Annals of Science* 45 (1988), pp. 113–152.

philosophy by John Theophilus Desaguliers, and this experience had stimulated him in the 1740s to appoint a physics lecturer (the Swiss scholar Samuel Koenig) at his own court. In those days, William also had ensured that a few craftsmen from Friesland and Groningen were trained as scientific instrument makers.[29] After the move to The Hague, Koenig continued his scientific lectures at the princely court. This apparently had a very stimulating effect on the elite and diplomatic "high society" in The Hague, resulting in a rather high number of scientific instrument cabinets. This also led to a curious side effect, like the The Hague lawyer Pieter Gabry, who, thanks to his ownership of a cabinet, the instruments of which he used to conduct meteorological and astronomical observations, managed to become a member of several prominent European scientific societies, despite the fact that he eventually was unmasked as a sciolist, publishing fake-observations as real.[30]

Use

About the actual use of the private cabinets of experimental philosophy in the Dutch Republic unfortunately almost nothing has been recorded. Even travel journals do not reveal much in this respect.[31] Based on the scarce available data, aspects such as competition and pleasure seem to have been at least as important as the results obtained. As a result, the scientific instrument was not only a tool for observation, demonstration, or experiment, it could also be used as a social vehicle.[32] Instruments could bring people together in amazement, physico-theological contemplation, competition or mere amusement (*Fig. 5*). In all these cases the collection contributed to the social prestige of its owner. Or, as Walters

29 Cf. H.J. Zuidervaart, *Speculatie, wetenschap en vernuft. Fysica en astronomie volgens Wytze Foppes Dongjuma (1707–1778), instrumentmaker te Leeuwarden*, Leeuwarden, 1995; Zuidervaart, "Reflecting 'popular culture': The Introduction, Diffusion and Construction of the Reflecting Telescope in the Netherlands", *Annals of Science* 61 (2004), pp. 407–452.

30 H.J. Zuidervaart, "A plague to the learned world: Pieter Gabry, F.R.S. (1715–1770) and his use of natural philosophy as a vehicle for gaining prestige and social status", *History of Science* 45 (2007), pp. 287–326.

31 Gerhard W. Kernkamp, "Bengt Ferrner's dagboek van zijne reis door Nederland in 1759", *Bijdragen en mededelingen van het Historisch Genootschap* 31 (1910), pp. 314–509 (also published in Swedish by Sten G. Lindberg (ed.), *Resa i Europa 1758–1762*, Uppsala, 1956); Kurt Møller Pedersen, Peter de Clercq (eds.), *An Observer of Observatories. The Journal of Thomas Bugge's Tour of Germany, Holland and England in 1777*, Aarhus, 2010.

32 Cf. Zuidervaart, "A plague to the learned world…", op. cit. (n. 30).

Fig. 5. Two persons performing electrical experiments in a Dutch cabinet with scientific instruments (Frontispiece from Jan Esdré, *Inleiding tot de kennisse der Natuurlyke Wysbegeerte*, vol. 1, Leiden, 1782).

has expressed it: "Scientific Instruments acted also as convergence points for conversation, companionship and consumption".[33]

A good example is the Amsterdam mayor and governor of the Dutch East India Company (VOC), Gerard Aernout Hasselaer. In 1758 he wrote to a friend in the Dutch East Indies that "from my youth on, the mathematical and physical sciences always have been my primary interest, yes I can say they have been like a doll for me".[34] The word "doll", or plaything, is revealing and underscores the entertaining aspect of the cabinets of experimental philosophy. The wealthy Hasselaer could afford a substantial cabinet with devices which were widely known "by their perfection and great number".[35] Still, Hasselaer's biography reveals that he not only took pleasure in physics, but that he also made use of the knowledge he acquired. It is known that he proposed or supported many technical innovations in for instance shipbuilding, hydraulics and pottery. It certainly was no coincidence that in 1751 the Dutch translation of Desaguliers' three-volume textbook *De Natuurkunde uit Ondervindingen Opgemaakt* (Physics made up from experiences) was dedicated to this scientific patron.[36] No wonder that after Hasselaer's death, in 1766, the Franeker professor Jan Hendrik van Swinden was very anxious to acquire this instrument collection for the Frisian university.[37] Although this attempt failed, ten years later many items from the Hasselaer collection were bought by university and other "professionals".[38]

Institutional Cabinets and Physics Societies

The fact that in 1776 many bidders at the Hasselaer auction were active professionals who were attached to some institution, illustrates the rise

33 Walters, "Conversation pieces…", op. cit. (n. 10), pp. 137–138.

34 Gerard Aernout Hasselaer, letter, December 1758: Amsterdam, City archive, Family Archive Hasselaer, inv. nr. 66.

35 Tiemen Cocquyt, "The Hasselaer auction of 1776: the transmission of scientific instruments from the public to the academic sphere", *Rittenhouse: the Journal of the American Scientific Instrument Enterprise* 22 (2009), pp. 70–89.

36 John Theophilus Desaguliers, *De Natuurkunde uit Ondervindingen Opgemaakt*, Amsterdam, 1736–1751, 3 vols.; Tirion's dedication to Hasselaer was added to the last volume in 1751.

37 Willem Boele S. Boeles, *Frieslands Hoogeschool en het Rijks Athenaeum te Franeker*, Leeuwarden, 1878–1889, 2 vols.; v.1, p. 421.

38 [Coenraad Fibus Metz], *Catalogus van eene zeer schoone verzameling van natuurkundige instrumenten […], nagelaten door wylen den wel edele gestrengen heer mr. G. A. Hasselaar*, Utrecht, 1776 (annotated copy in Utrecht University Library).

of the institutional cabinet of experimental philosophy in the last quarter of the eighteenth century. One of the first institutional cabinets (apart from the one at Leiden University) was formed in 1761 at the Doopsgezinde Kweekschool (the "Mennonite Seminary") in Amsterdam.[39] That an institution for the theological education of preachers needed a rather large cabinet of experimental philosophy, is in itself a beautiful testimony of the importance attached at the time to physico-theology. Other educational institutions, such as the three Renswoude foundations in Delft, The Hague and Utrecht, as well as the Leiden Society *Mathesis Scientiarum Genitrix*, also collected an impressive cabinet with scientific instruments, which in those cases were mostly used for the technical training of mathematically gifted orphans.[40]

In the last quarter of the eighteenth century, this tendency toward institutionalisation grew enormously. In the Dutch Republic many general oriented learned societies were established, some of them with a special branch or focus on (applied) physics, such as in Rotterdam, the Bataafsch Genootschap der Proefondervindelijke Wijsbegeerte (Batavian Society for Experimental Philosophy), in Haarlem the Teylers Genootschap (Teyler Foundation) and in Amsterdam the *Felix Meritis* Society. Also, in almost all the cities of the country local physics societies were founded (see Graph 2, *Fig. 6*).

These physics societies almost all brought together a collection of scientific instruments. As a rule, these instruments were employed only for instruction and demonstration purposes, but sometimes members were allowed to use the instruments for research. However, innovative research was conducted only occasionally. For instance, the groundbreaking physico-chemical and electrical experiments of the small Gezelschap der Hollandsche Scheikundigen (Society of Dutch Chemists), were performed in the private sphere, although most of the experimenters were members of the *Felix Meritis* Society. This is rather remarkable, because the brand new

39 H.J. Zuidervaart, " 'Meest alle van best mahoniehout vervaardigd'. Het kabinet van filosofische instrumenten van de Doopsgezinde Kweekschool te Amsterdam, 1761–1828", *Gewina. Tijdschrift voor de Geschiedenis der Geneeskunde, Natuurwetenschappen, Wiskunde en Techniek* 29 (2006), pp. 81–112. Reprinted in: *Doopsgezinde Bijdragen. Nieuwe reeks* 34 (2008), pp. 63–104.

40 L. Roberts, "Instruments of science and citizenship: Science education for Dutch orphans during the late eighteenth century", *Science & Education. Contributions from History, Philosophy and Sociology of Science and Mathematics* 21 (2010), pp. 157–177. See also: Engelina P. de Booy, Johannes Engel, *Van erfenis tot studiebeurs. De Fundatie van Renswoude te Delft. Opleiding van wezen tot de 'vrije kunsten' in de 18e en 19e eeuw. De fundatiehuizen. Bursalen in deze eeuw*, Delft, 1985; Carin Gaemers, *Nalatenschap als toekomst. De fundaties van de Vrijvrouwe van Renswoude 1754–1810*, Zutphen, 2004.

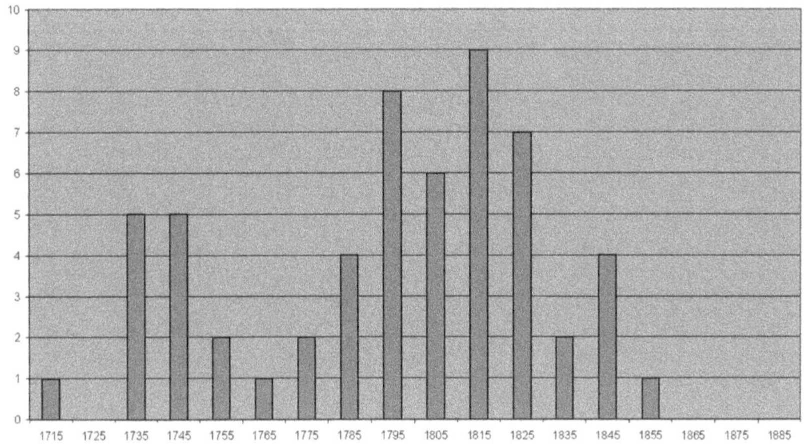

Fig. 6. Graph 2: Dates of the founding of local physics societies in the Northern Nederlands.

Felix Meritis building (established 1789) was equipped with an up-to-date chemical laboratory, specifically designed for research purposes.[41]

The Composition of the Cabinets

In terms of composition the institutional cabinets initially had their emphasis on mechanics and pneumatics, while in the nineteenth century the emphasis shifted towards electricity (*see Table 3*). This contrasts with the larger private cabinets, which in the eighteenth century had their emphasis mostly on optics, although it should be noted that this could diverge according to the taste of each individual owner. After the turn of the nineteenth century the instrument cabinet seems to become less elitist. Around 1800 there are also ordinary shopkeepers who possess scientific instruments, although the larger collections still are owned by professionally active teachers and wealthy aristocrats. This egalitarian development probably was boosted by the steep decline in prices after the revolutionary years of 1787 and 1795. For example, an exquisite planetarium from Hasselaer's cabinet, sold in 1776 for more than 1.300 guilders, was auctioned in 1800 for only 200 guilders.

[41] Harry A.M. Snelders, *Het Gezelschap der Hollandsche Scheikundigen: Amsterdamse Chemici uit het Einde van de Achttiende Eeuw*, Amsterdam, 1980.

Table 3. Six Institutional Instrument Cabinets.[42]

	1789	1802	1815	1828	1859	1871		
	Renswoude Foundations		*Diligentia* Society	Mennonite Seminary	Teyler Museum	*Felix Meritis* Society	Cumulative	
	Delft	The Hague	The Hague	Amsterdam	Haarlem	Amsterdam	No.	%
Mechanics	*38*	*135*	*103*	*49*	25	78	428	20
Electricity	35	48	96	7	*83*	*119*	388	18
Pneumatics	38	72	86	48	25	79	348	16
Optics	31	88	27	36	42	66	290	14
Hydraulics	22	65	31	28	18	33	197	9
Heat	19	19	15	2	16	33	104	5
Mathematics	20	36	18	3	6	0	83	4
Astronomy	18	14	15	4	4	22	77	4
Models	3	34	4	0	19	0	60	3
Miscellaneous	0	0	35	2	14	0	51	2
Magnetism	0	12	6	2	10	16	46	2
Sound	0	8	0	1	11	21	41	2
Artificial airs	0	0	23	0	7	0	30	1
Total	224	531	459	182	280	467	2.143	100

Another example of the egalitarian tendency around 1800 is that the phenomenon of the local physics society now also emerged in rural areas. In the first decades of the nineteenth century one could find almost no place in the Netherlands, where there was not a local physics society with its own instrument cabinet. From Wormerveer to Sneek, or from Beers

42 Only the number of entries in the inventories have been counted. In some cases a number can refer to a few related items. Used are the following inventories: [1] Renswoude Delft (1789): Catalogue printed in: De Booy, Engel, *Van erfenis tot studiebeurs…*, op. cit. (n. 40); [2] Renswoude Den Haag (1802), See: [Jacob Florijn], *Naamlijst en Korte Beschrijving van alle de Wis- en Natuurkundige Werktuigen, behoorende aan de Stichting van Wylen de Hoog Edele Welgeboorene Vrouwe, Vryvrouwe van Renswoude, in het Haagsche Burger Weeshuis*, The Hague, 1802; [3] Diligentia (1815): "Naamlijst en beschrijving der Natuurkundige Werktuigen behoorende aan de Maatschappij Diligentia in Den Haag": Leiden, Museum Boerhaave; see also: Peter Wisse, "The Philosophical Society Diligentia [The Hague] and its instrument collection", *Bulletin of the Scientific Instrument Society*, no. 64 (2000), pp. 3–8; [4] "Doopsgezinde Kweekschool" (1828): *Catalogus van eene uitmuntende verzameling Optische, Phijsische, Mathematische en andere Instrumenten* […], *Al hetwelk verkocht zal worden op Dinsdag den 23 December 1828 en volgende dagen*, Amsterdam, 1828 (copy in Archive of the "Verenigde Doopsgezinde Gemeente van Amsterdam" in the City Archive Amsterdam, inv. nr. 1376); [5] Teylers Museum (1859): "Catalogus der natuurkundige werktuigen…", op. cit. (n. 8); [6] *Felix Meritis* (1871): "Catalogus van het Physisch Kabinet…", op. cit. (n. 2).

to Meppel, local physics societies were founded.[43] Education – particularly to young people – seemed to have been the main objective of these nineteenth-century institutions.[44] Until the reform of the Dutch educational system, in 1863, these physics societies played a crucial role in the dissemination of current scientific and technical knowledge.

The Downfall of the Cabinet of Experimental Philosophy

As is illustrated by Graph 1 (*Fig. 3*), after c. 1870 the scientific instrument collections disappeared rather suddenly from the scene. This abrupt decline probably was related to two issues. First, in the mid-nineteenth century, physics became both more difficult and more professionalized, and on the other hand the far-reaching education reform of 1863 suddenly had put physics on the school curriculum.[45] The instructive role of the various physics societies simply had become obsolete. Thus most physics societies abolished their instructors and disposed of their instruments.[46] Some societies were discontinued; others converted their activities into the reading of papers. The official responsible for the implementation of the new *Dutch Education Act*, Daniël Jan Steyn Parvé, recognized the opportunity of this development. On his initiative, in the 1860's and 1870's various newly created *Hogere Burger Scholen* (Senior Citizen Schools) made agreements with local physics societies. As a result of this initiative, many scientific instruments ended their lives (and were worn out) in one of the new secondary schools.

---

43 H.A.M. Snelders, "De natuurwetenschappen in de lokale wetenschappelijke genootschappen uit de eerste helft van de negentiende eeuw", *De negentiende eeuw* 7 (1983), pp. 102–122.

44 A nice example of the educational aspect is the *Natuurkundig Schoolboek* by Johannes Buys, published in 1798 by the "Maatschappij tot nut van 't Algemeen". This book was reprinted many times. Its author, Johannes Buys, was Lector in Physics at the *Felix Meritis* Society from 1795 until 1818.

45 P. de Clercq, "The scientific instrument-making industry in the Netherlands in the nineteenth century", in de Clercq (ed.), *Nineteenth-century scientific instruments and their makers*, Leiden, 1985; pp. 205–225: 220.

46 See for instance: *Catalogus van eene fraaije verzameling instrumenten en toestellen voor wis-, werktuig- en natuurkunde toebehorende aan het Natuurkundig Genootschap te Groningen, welke publiek zal worden verkocht op zaterdag 7 januari 1865*, Groningen, 1865. That as late as 1880 the Physics Society in Leeuwarden released a printed catalogue of its instrument collection, is evidently an exception. *Natuurkundige instrumenten en andere eigendommen van het Natuurkundig Genootschap te Leeuwarden*, Leeuwarden, 1880.

The Cabinet of Experimental Philosophy as a Precursor of the Physics Laboratory

The disappearance of the cabinet of experimental philosophy raises two questions: first, what historical meaning can be given to these eighteenth- and ninetenth-century cabinets; second, why did only such a small percentage of the instruments that built these cabinets survive until today?

To start with the first question: rather recently, in 2005, Maurice Crosland has proposed an important legacy of the *cabinet de physique*. He claims that the cabinet of experimental philosophy has served as the breeding ground for the modern physics laboratory.[47] This *Natuurkundig Laboratorium* (or "NatLab") is still rather young – centuries younger than, for example, the chemical laboratory. These chemical workplaces, with their furnaces and retorts, can be traced back to the medieval alchemical investigations. However, the nineteenth-century physics laboratory has a very different genesis. According to Crosland the "NatLab" evolved directly from the eighteenth-century instrument cabinet. Can this interesting hypothesis be confirmed for the Netherlands?

Indeed we are able to illustrate this very development, especially at the Teyler Museum in Haarlem and the Natuurkundig Gezelschap in Utrecht. Already in 1790 Van Marum, the director of Teyler's cabinet of experimental philosophy, succeeded in purchasing two houses next to the – still present – Oval Room (*Pl. I*). Here he installed a research facility, then already coined a *Laboratorium*, where he performed several physical and chemical experiments, for instance with his large electrical machine.[48] According to a contemporary journal this facility was designed to "promote the knowledge of nature, especially for doing those investigations in physics, which for most physicists are too expensive or laborious to take upon at their own account".[49]

---

[47] Maurice Crosland, "Early Laboratories c. 1600–c. 1800 and the Location of Experimental Science", *Annals of Science* 62 (2005), pp. 233–253.

[48] Martinus van Marum, "Brief aan den Heer Jan Ingenhousz... inhoudende ene beschryvinge van een Electriseer-Machine", *Algemene konst- en letter-bode* 2:1 (1791), 188.

[49] "De Teylersche Stigting", *Algemene konst- en letter-bode* 2:2 (1791), p. 210: "er is ook kortelings, door het aantrekken en vertimmeren van twee naast elkanderen gelegen woningen, een ruim en ongemeen wel ingerigt *Laboratorium* aan het Stigtingshuis gehecht, en met den nodigen toestel tot Physische en Chemische Proefnemingen en nasporingen voorzien geworden. [...] ter bevordering van natuurkennis, inzonderheid zodanige Physische nasporingen te doen, als voor de meeste natuurkundigen te kostbaar of te omslagtig zyn om voor hunne eigene rekening ondernomen te worden".

This research laboratory stayed operational until 1884, when a modernized *Natuurkundig Laboratorium* at the Teyler Museum opened its doors.[50] This facility (of which the Nobel laureate Hendrik Antoon Lorentz was the curator from 1910 to 1928) eventually was dissolved in 1955.

In Utrecht we can observe an interesting variant of this same development. Here a physics society was established in 1777, at the initiative of Johannes Rossijn, Professor of Philosophy at Utrecht University. The year before, he was one of the buyers at the auction of the aforementioned instrument cabinet of Gerard Aernout Hasselaer. Rossijn bought 25 instruments for the University, whose cabinet possessed at that time only 66 instruments. So this cabinet was expanded considerably. For his acquisition budget Rossijn was fully dependent of the Utrecht city government, who financed the University in its entirety. However, exactly in those years the city of Utrecht experienced a major financial crisis. That at this very moment Rossijn had been granted an acquisition budget of 600 guilders for the Hasselaer collection, was very exceptional. It was the highest amount ever granted in Utrecht for the purchase of new scientific instruments. However, this was all that could be hoped for. In order to assure himself of an on-going budget for the purchase of scientific instruments, Rossijn needed "external funding". To obtain this, he proceeded to an unorthodox method.

On the Hasselaer auction, Rossijn had carefully noted the wealthy buyers from the Utrecht aristocracy. He probably knew that in 1761 the instrument cabinet of the Mennonite Society in Amsterdam had been fully financed by private lenders.[51] So why not replicate this initiative? Thus, half a year after the auction, he proposed to the most prominent buyers to found together a Gezelschap ter Beoeffening en Bevordering van de Proef-ondervindelijke Natuurkunde (Society for the Practise and Promotion of Experimental Physics). The first co-founder was even Hasselaer's grandson and namesake, Gerard Aernout Taets van Amerongen, an immensely rich nobleman.

---

50 Marijn van Hoorn, "Elisa van der Ven and the Physical Laboratory of the Teyler Foundation (Haarlem) 1878–1909", in Robert G.W. Anderson, Jim A. Bennett, William F. Ryan (eds.), *Making Instruments Count: Essays on Historical Scientific Instruments presented to Gerard L'Estrange Turner*, Aldershot, 1993; pp. 278–290; Gerard L'E. Turner, "Teyler's Museum, Haarlem, during the nineteenth century", in P. de Clercq (ed.), *Nineteenth-century scientific instruments and their makers*, Leiden, 1985; pp. 227–240: 238.

51 Zuidervaart, "Meest alle van best mahoniehout vervaardigd…", op. cit. (n. 39), appendix 1.

It was a clever step to take. Far into the nineteenth century the physics society would finance most of the equipment used by the University's physics department.[52] In 1818 the University and the Society financed together a more suitable location for the physics department. Pieter Hendrik van Cittert, the first director of the Utrecht University Museum, called this location a "Physics Laboratory", but we have not been able to confirm a contemporary use of this term. In any case, what Van Cittert remarked in 1929 is reasonably correct:

> The ties that connected the Physics Society to the University had become so tight, that it is difficult to distinguish the scientific achievements of the Laboratory from those of the Society. An example is the famous determination of the speed of sound, conducted in 1823 by Professor G.D. Moll and Dr. A. van Beek, the treasurer of the Physics Society, with instruments which were funded half by the University and half by the Society.[53]

In 1835 it was stipulated that the Utrecht University instrument cabinet was the best equipped of the country, excelling those of all other universities and institutions in the Netherlands, this all as the result of the ongoing support of the "flourishing Physics Society here".[54] That this statement was no exaggeration, is demonstrated by an inventory of 1838, which lists no fewer than 1,239 scientific instruments, jointly owned by the Utrecht Physics Society and the University.[55] This rich cabinet eventually formed the germ of the new physics laboratory of Utrecht University, founded in 1877. Only eight years before, the financial contribution of the Physics Society to the University's physics department had stopped. Finally, in 1889, all instruments of the Physics Society were donated to the laboratory.[56]

---

[52] Johanna Geertruida van Cittert-Eymers, "Het Natuurkundig Gezelschap te Utrecht 1777–1977", in *NG 200. Natuurkundig Gezelschap te Utrecht 1777–1977*, Utrecht, 1977, pp. 39–82.

[53] Pieter Hendrik van Cittert, "Geschiedenis van de verzameling antieke instrumenten van het Natuurkundig Laboratorium der Rijks Universiteit en van het Natuurkundig Gezelschap", *Jaarboek Oud-Utrecht* (1929), pp. 1–27: 23.

[54] *Tijdschrift voor Geschiedenis, Oudheden, Merkwaardige Bijzonderheden en Statistiek van Utrecht* 1 (1835), p. 255.

[55] Manuscript inventory at the Utrecht University Museum. See more in detail: Johannes MacLean, "De verwerving van de natuurwetenschappelijke collecties aan de Utrechtse Universiteit van 1815–1848", *Jaarboek Oud-Utrecht* (1975), pp. 73–98: 82–83.

[56] Van Cittert-Eymers, "Het Natuurkundig Gezelschap te Utrecht…", op. cit. (n. 52), p. 66.

But how about the genesis of the physics laboratory of Amsterdam University? For we have seen that in that same year 1889 most of the scientific instruments that the Amsterdam Physics Laboratory had received from the *Felix Meritis* Society were sold. Nonetheless it can be argued that also in Amsterdam, the University's "NatLab" found its genesis in a private instrument cabinet. Amsterdam University, established in 1877, was a direct successor to the *Athenaeum Illustre*, founded in 1632. Courses in experimental philosophy were already given in the 1780s by the well-known professor Jan Hendrik van Swinden, but the instruments he used were those of the Mennonite Seminary.[57] It was not before 1847 that the *Athenaeum* acquired its first scientific instruments, when this institution purchased the instrument cabinet of the deceased Professor of Philosophy, Willem Simon Swart.[58] Thus in Amsterdam Swart's instrument cabinet was the stepping stone to the Amsterdam Physics Laboratory, for which a new building was erected in 1880. So, here too, Crosland's observation appears to be correct.

The Scientific Instrument as a Museum Object

Then the last question: why did only such a small percentage of the instruments that built the cabinets of experimental philosophy survive until today? What is the reason that most instruments evidently were sold for scrap?

Unfortunately, it is like today: obsolete scientific instruments are rarely seen as being of historic importance, worthy to be preserved in their own right. This is completely different from objects of art. Although – in a few isolated cases – the notion that scientific instruments can testify to an interesting past becomes noticeable during the nineteenth century,[59]

---

[57] Zuidervaart, " 'Meest alle van best mahoniehout vervaardigd'...", op. cit. (n. 39), pp. 97–98.

[58] At the start of his professorate, in 1834, Swart had received an allowance for a private purchase of scientific instruments. His cabinet was bought by the *Athenaeum* in 1847 for 2500 guilders. Cf. *Van Athenaeum tot Universiteit...*, op. cit. (n. 2), pp. 63–64.

[59] Anthony J. Turner, "From Mathematical Practice to the History of Science. The Pattern of collecting Scientific Instruments", in P. de Clercq, A.J. Turner (eds.), *Origins and Evolution of Collecting Scientific Instruments,* Oxford 1995 [Special issue of the *Journal of the History of Collections* 7 (1995), 135–150]; Turner, "The collecting, trade and display of early scientific instruments, 1830–1930", in de Clercq (ed.), *Scientific Instruments: Origins and Imitations. The Mensing Collection,* Leiden, 1999; 23–48.

most cabinets of experimental philosophy were simply dissolved and cannibalized. The fate of most surviving instruments is a textbook example of what Michael Thompson has described in his *Rubish Theory: the Creation and Destruction of Value* (1979). In this book Thompson argues that obsolete objects can regain a value only when they first are dumped from their original function, regarded as being superseded and worthless.[60] This is exactly what has happened with most scientific instruments, after they had fallen into disgrace. In the decades before and after the turn of the twentieth century many of these instruments were destroyed. What ultimately survived has been preserved mostly thanks to happy contingencies.

Again, the collection of the Utrecht Physics Society offers a nice example. How these instruments have survived is vividly chronicled by Mrs. J.G. van Cittert-Eymers, who had learned the state-of-affairs first-hand from her late husband, Pieter Hendrik van Cittert:

> On 19 December 1917 the last resident of the Utrecht physics laboratory, the amanuensis Mr. M.F. Filbri died. So, his widow and daughters were forced to leave the rooms that they had used, which space then became available for the laboratory. The daughter's bedroom had been in the attic of the building and because of the current ideas about 'what is tasteful', never before had any laboratory employee been allowed access to this attic. After the departure of the Fibri family, Dr. P.H. van Cittert – in 1912 an assistant in physics – entered the attic and to his astonishment he found there many very old instruments: approximately some 1,000 pieces [...]. Before that moment no one had any idea about their existence and origin. It appeared, from labels at the instruments, that over 40% had come from a – then forgotten – Physics Society.[61]

This discovery led in 1928 to the revitalisation of the – nowadays still active – *Natuurkundig Gezelschap Utrecht*, after the "Physics Society" had become dormant in 1892. Finally – after years of deliberation – it resulted in 1936 to the creation of the Utrecht University Museum, shortly after an exhibition which Van Cittert had organized to commemorate the third centenary of Utrecht University.[62]

---

[60] Michael Thompson, *Rubish Theory. The Creation and Destruction of Value*, Oxford, 1979.

[61] Van Cittert-Eymers, "Het Natuurkundig Gezelschap te Utrecht...", op. cit. (n. 52), p. 71.

[62] Cf. P.H. van Cittert, *Tentoonstelling van natuurkundige instrumenten, ten behoeve van het "Theatrum Physicum" aangekocht in het tijdvak 1706–1776 en ten behoeve van het "Natuurkundig Gezelschap" in het tijdvak 1777–1816*, Utrecht, 1931.

A similar story can be told for Leiden. Here too, the abandoned scientific instruments had been expelled to a laboratory attic.[63] In 1907 these instruments were allowed to re-emerge for a short time for an exhibition during the eleventh *Nederlandsch Natuur- en Geneeskundig Congres* (Dutch Natural and Medical Congress), but after that they were quickly sent back to their attic.[64] Starting in 1916 these orphaned instruments were studied as historical items by the physicist Claude August Crommelin.[65] After years of hard labour, in 1928 he managed to found a museum around these instruments, of which institution he became the first director. This Nederlandsch Historisch Natuurwetenschappelijk Museum (Netherlands Historical Natural Science Museum), opened its doors to the public in 1931.[66] So in this way the "rubbish" eventually had received a new meaning and a different – historical – appreciation. Only at the Haarlem Teyler Museum, this changing meaning was achieved gradually in the original setting of the instruments, but this story is told elsewhere in this volume.[67]

Conclusions

The cabinet of experimental philosophy in the Netherlands emerges in the first decade of the eighteenth century and disappears in the last decade of the nineteenth century. Institutional collections of scientific instruments were formed by the end of the eighteenth century. Besides entertainment, contemplation and sociability, education seemed a prominent motive for the founding of a cabinet of experimental philosophy, especially in the local physics societies. The cabinets with scientific instruments created a significant market for instrument makers and dealers. This development,

---

63 De Clercq, *The Leiden Cabinet of Physics*…, op. cit. (n. 17), p. 13.

64 Evert C. van Leersum, François M.G. de Feyfer, Philip C. Molhuysen, *Catalogus van de geschiedkundige tentoonstelling van natuur- en geneeskunde te houden te Leiden 27 Maart–10 April 1907, ter gelegenheid van het Elfde Nederlandsch Natuur- en Geneeskundig Congres*, Leiden, 1907.

65 Claude August Crommelin, *Beschrijvende catalogus der historische verzameling van natuurkundige instrumenten in het Natuurkundig Laboratorium der Rijks-Universiteit te Leiden*, Leiden, 1926; Crommelin, *Catalogus van de historische tentoonstelling van astronomische instrumenten*, Leiden, 1928.

66 The "Nederlandsch Historisch Natuurwetenschappelijk Museum" became the "Rijksmuseum voor de geschiedenis der Natuurwetenschappen", which was transferred into the present "Museum Boerhaave". Cf. C.A. Crommelin, "Het Nederlandsch Historisch Natuurwetenschappelijk Museum", *Oudheidkundig Jaarboek* 10 (1930), pp. 10–25.

67 Martin Weiss, "Monuments of Science. How the Teyler Museum's Instrument Collection became Historical", *infra*, pp. 195 ff.

added to the scientific instruction given in the numerous physics societies, must have contributed significantly to the formation of technical expertise in the Netherlands. Further, the Dutch situation supports Crosland's hypothesis that several physics laboratories emanate from the eighteenth-century *Cabinet de Physique*.

The scientific instruments from these cabinets were mostly dismantled over time. Only in a few documented cases more-or-less complete collections have survived after a period of "Rubbish". In the *inter-bellum* period of the twentieth century, some historically oriented (mostly university) museums were founded, where surviving historic scientific instruments have found a new home and new meaning.[68]

[68] See about the general European development with regard to the founding of university museums: Marta C. Lourenço, *Between Two Worlds. The Distinct Nature and Contemporary Significance of University Museums and Collections in Europe*, Paris, 2005, p. 80.

ENTREPRENEURS IN EXPERIMENTS: THE LEIDEN CABINET OF PHYSICS AND THE MOTIVES OF ITS FOUNDERS (1675–1742)

Hans Hooijmaijers and Ad Maas

Introduction

This article concerns the Leiden Cabinet of Physics – the corpus of instruments used for the instruction of natural philosophy at Leiden University – from its foundation in 1675 until the death of its most famous director Willem Jacob 's Gravesande in 1742. The Cabinet was founded by professor Burchard de Volder after a period in which natural philosophy at Leiden University was hopelessly entangled by heated metaphysical and religious arguments.[1] De Volder's source of inspiration had been the Royal Society in London, where the experimental method among other things had been instrumental in detaching natural philosophy from the religious and political arguments which deeply divided British society in those days. Scientific instruments – the air-pump especially became the paragon of this approach – yielded experimentally produced irrefutable "matters of fact" capable of distracting attention from bitterly contested dogmatic arguments.[2]

From 1705, when de Volder retired, until 1724, Wolferd Senguerd (1646–1724), extraordinary professor in peripatetic philosophy was the director of the *Theatrum Physicum* (as the place was named where the Cabinet was stored and the experimental courses were conducted). Yet, he did not really seem to have used the theatre in an innovative manner and, for that matter, taught private lessons at his home where he had his own cabinet. Senguerd was eclipsed, from 1717 onwards, by 's Gravesande, who started to acquire world-fame with his experimental private courses, in which he

---

1 Edward G. Ruestow, *Physics at 17th and 18th Century Leiden: Philosophy and the New Science in the University*, The Hague, 1973; Gerhard Wiesenfeldt, *Leerer Raum in Minervas Haus: Experimentelle Naturlehre an der Universität Leiden, 1675–1715*, Amsterdam, 2002; Adriaan C. De Hoog, *Some Currents of Thought in Dutch Natural Philosophy: 1675–1720* (PhD-thesis), Oxford, 1974.

2 Steven Shapin, Simon Schaffer, *Leviathan and the Air-Pump: Hobbes, Boyle, and the Experimental Life*, Princeton, 1985.

disseminated the ideas of Isaac Newton with his ingeniously constructed instruments. By attracting numerous students from abroad and by writing the first Newtonian physics handbook (in which his instruments figured prominently) he contributed considerably to the dissemination of Newton's natural philosophy all over Europe. According to Jonathan Israel, 's Gravesande was even one of the main initiators of Anglomania – the absorbing hunger for English ideas and achievements in Europe particularly in the 1730s and 1740s.[3] After Senguerd's death in 1724 's Gravesande became in charge of the *Theatrum Physicum*.

On the surface, then, the Leiden Cabinet of Physics was a collection of instruments with which British methods and ideas were translated into the context of a (Dutch) university. From there, consequently, these were disseminated further over the Continent. In this article however, we will choose another approach than the history of ideas. We will consider the development of the Cabinet from the perspective of the local needs of Leiden University and the personal interests and preferences of its directors. This perspective will reveal other motives that have been involved and shed another light on the influence of British philosophy.

As we will argue, the fact that the Leiden Cabinet of Physics was tailored to local needs and ideas is clearly reflected in the instruments themselves. We will present a number of well-known instruments of the Leiden Cabinet of Physics, which reveal the multiple motives and sources of inspiration of the illustrious founding fathers. We do not pretend, by the examples we elaborate, to be able to present a complete survey of all motives that have possibly driven de Volder and 's Gravesande, which prove to be, incidentally, sometimes hard to reveal. Before looking at the instruments and their stories, we will start with the trips to England made by de Volder and 's Gravesande, both of which determined the development of the Cabinet in a decisive manner.

Two Trips to England

From the 1640s onwards, the Dutch universities had been afflicted by religious and philosophical controversies, with orthodox Calvinists confronting their more liberal fellow believers. Roughly along the same divide, scholastics opposed Cartesians. For Leiden University especially the years

[3] Jonathan Israel, *Radical Enlightenment: Philosophy and the Making of Modernity 1650–1750*, Oxford, 2001, pp. 555–568.

1672–1673 had been troublesome. Medicine and natural philosophy in particular became battle grounds in which Cartesians and anti-Cartesians fought out bitter arguments. According to Adriaan Cornelis de Hoog and Gerhardt Wiesenfeldt, the main motive for de Volder to plead for a cabinet of physics was to end these heated metaphysical arguments.[4]

Burchard de Volder (1643–1709) had studied philosophy in Amsterdam (at the Illustrious School) and at Utrecht University, where Johannes de Bruyn taught him about the new mechanistic philosophy.[5] After taking his doctorate he left for Leiden to study medicine, where he was influenced by Fransiscus de le Boë Sylvius (1614–1672), who combined iatrochemical and mechanistic explanations in medicine. Belonging as he did to the Mennonite religious denomination, de Volder took practice as a "poor doctor" in the Remonstrant community of Amsterdam. The Remonstrants were a secession of the Dutch Reformed Church that rejected the doctrine of predestination and in general took a more liberal attitude in both religious as well as earthly matters. In the Amsterdam Remonstrant community there was a lively interest in new philosophical and religious ideas. Perhaps it was this progressive environment that further stimulated de Volder's predilection for new, mechanistic natural philosophical theories, and his bent for Cartesian thoughts in particular came to fruition. De Volder made acquaintance with leading scholars like the theologian Philipp van Limborch (1633–1712), with whom he corresponded about religious tolerance, and Johannes Hudde (1628–1704), an influential regent (and accomplished mathematician) who used his network to have de Volder appointed as philosophy professor in Leiden.

Leiden University proved to be a less tolerant environment than the Amsterdam Remonstrant community, and de Volder soon found himself, as a Cartesian, involved in the quarrels that split the academic community in two. Yet de Volder himself was anything but a quibbler in philosophical matters. According to de Hoog and Wiesenfeldt he became tired of the continuing metaphysical controversy, and also increasingly critical of some of René Descartes' views. He decided to seek refuge in the new experimental natural philosophy coming from Britain, where it appeared

[4] De Hoog, op. cit. (n. 1), p. 143; Wiesenfeldt, op. cit. (n. 1), pp. 7, 53 and 111–112.

[5] Biographical information of the Volder and the foundation of the *Theatrum Physicum* can be found in Jean Le Clerc, "Eloge de feu Mr. de Volder professeur en philosophie et aux mathematiques, dans l'academie de Leide", *Bibliotheque choisie, pour servir de suite a la Bibliotheque universelle* 18 (1709), pp. 346–401; Jacob Grovonius, *Burchardi de Volder laudation*, Leiden, 1709; De Hoog, *Some currents…*, cit. (n. 1), pp. 122–185; Wiesenfeldt, op. cit. (n. 1), pp. 41–64.

to be successful in detaching natural philosophy from the metaphysical and religious sensitivities.[6]

On a trip to England in 1674 he visited the British scholar Isaac Newton in Cambridge and the Royal Society in London, where he might have met Boyle.[7] Possibly his contacts in Amsterdam yielded him an introduction. At the Royal Society he could learn how scientific instruments could be deployed to produce irrefutable "matters of fact" capable of superseding bitterly contested dogmatic arguments. Historians conclude that de Volder realized that scientific instruments could likewise be used to get Leiden University out of hot water.[8] Little is known about his stay in England, yet one year later de Volder would make an appeal for his *Theatrum Physicum* to the board of the university, which responded quickly and positively. De Volder was given a building for the theatre adjacent to the botanical gardens. The theatre was built in the way of an anatomical theatre, with banks in a circle around an erased platform in the middle (unfortunately only an image of the outside survived, *Pl. II*).[9] Furthermore he was granted four hundred guilders to spend on its equipment. His experimental lessons started in 1676.

In 1715 the young lawyer Willem Jacob 's Gravesande (1688–1742) travelled for a completely different reason to England. He was born in 's-Hertogenbosch, a town in the Catholic south of the Dutch Republic. Yet the 's Gravesande – or Storm van 's Gravesande as the full family name was – were no Catholics, but belonged to the Protestant administrative upper echelon of the town.[10] As a patrician's son, Willem Jacob

---

[6] De Hoog, op. cit. (n. 1), p. 143; Wiesenfeldt, op. cit. (n. 1), pp. 7, 53 and 111–112.

[7] Willem Otterspeer, *De vesting van de macht: De Leidse universiteit, 1673–1775. Groepsportret met dame II*, Amsterdam, 2005, p. 55; Wiesenfeldt, op. cit. (n. 1), p. 61.

[8] De Hoog, op. cit. (n. 1), p. 143; Wiesenfeldt, op. cit. (n. 1), pp. 7 and 111–112; Otterspeer, op. cit. (n. 7), pp. 54 and 108.

[9] Zacharias C. Von Uffenbach, *Merkwürdige Reisen durch Niedersachsen, Holland und Engelland*, Frankfurt, 1753–1754, pp. 425 and 454.

[10] The main source of 's Gravesande's biography is the account written by Jean N.S. Allamand, "Histoire de la vie et des ouvrages de Mr. 's Gravesande", in Allamand (ed.), *Oeuvres philosophiques et mathématiques de Mr. G.J. 's Gravesande*, Amsterdam, 1774, pp. ix–lix. See also Allamand, "'s Gravesande", in P. Marchand, *Dictionaire historique ou mémoires critiques et littéraires*, The Hague, 1758–1759, 2 vols.; v. 2, pp. 214–242: 224–227. Jean N.S. Allamand (1713–1787) had been a confidant of 's Gravesande and eventually became a successor as professor of mathematics and philosophy in Leiden. Secondary literature concerning 's Gravesande: Pieter L. Rijke, "Levensschets van Willem Jacob 's Gravesande", *Album der natuur: Een werk ter verspreiding van natuurkennis onder beschaafde lezers van allerlei stand*, new series 27 (1879), pp. 65–88; Cees De Pater, "Willem J. 's Gravesande", in A.J. Kox (ed.), *Van Stevin tot Lorentz: Portretten van achttien Nederlandse natuurwetenschappers*, Amsterdam, 1990, pp. 81–92; De Pater (ed.), *Willem Jacob*

was educated by a private teacher called Isaac Tourton. According to 's Gravesande's biographer Allamand (whose description does not appear reliable in every respect), Tourton's lessons encouraged 's Gravesande's talents and interest in mathematics. Together with two of his brothers, however, Willem Jacob was destined for the law course at Leiden University. He studied law from 1704 to 1707, during which period he is also said to have written his first mathematical treatise, *Essai de perspective*. This, however, was not published until 1711.[11]

's Gravesande did not receive a degree from the Faculty of Philosophy. There is also no evidence that he ever attended the courses and demonstrations of the philosophy professors de Volder or Senguerd; in any case he appears never to have defended a philosophical or mathematical disputation under their direction.[12] 's Gravesande probably was not considering an academic career in this field by this time. His correspondence with the Swiss mathematician Nicolaus (I) Bernoulli (1687–1759) and the physician and amateur mathematician Bernard Nieuwentijt (1654–1718) shows that (at least in the years after his study) he nevertheless still engaged in mathematics.[13] After finishing his thesis on suicide – he maintained that it was a reprehensible deed – he set up practice (with his brothers) as a barrister in The Hague.[14]

In 1715 the States General sent the influential nobleman Arent Wassenaer van Duyvenvoorde (1669–1721) to England as a special envoy to congratulate George I on his accession to the throne and to carry on some tough diplomatic negotiations. Wassenaer van Duyvenvoorde decided to take the young lawyer Willem Jacob 's Gravesande with him as his "first secretary". In addition, 's Gravesande acted as one of the private teachers of Duyvenvoorde's son Brilanus. The other teachers were 's Gravesande acquaintance Justus van Effen (1684–1735) and no less a figure than John Theophilus Desaguliers (1683–1744), who became friends with 's Gravesande

's Gravesande, welzijn, wetenschap en wijsbegeerte, Baarn, 1998, pp. 23–58; Ad Maas, "The Man Who Erased Himself: Willem Jacob 's Gravesande and the Enlightenment", in Eric Jorink, Ad Maas, *Newton and The Netherlands: How Isaac Newton was Fashioned in the Dutch Republic*, Leiden, 2012, pp. 113–137.

11 Allamand, *Oeuvres philosophiques*..., cit. (n. 10), pp. ix–xi.

12 Wiesenfeldt, op. cit. (n. 1), p. 248.

13 Eddie Shoesmith, "The Continental Controversy over Arbuthnot's Argument for Divine Providence", *Historia Mathematica* 14 (1987), pp. 133–146; Rienk Vermij, *Secularisering en natuurwetenschap in de zeventiende en achttiende eeuw: Bernard Nieuwentijt*, Amsterdam, 1991, pp. 118–120.

14 Willem J. 's Gravesande, *Dissertatio juridica inauguralis de autocheiria*, Leiden, 1707.

and later translated his physics handbook into English.[15] A few months after arriving in England, 's Gravesande was elected a fellow of the Royal Society. It was William Burnet (1687–1729), an old university friend, who introduced him. Here 's Gravesande met President Isaac Newton in person. Unfortunately, no account exists of this meeting.

Allamand has claimed that the personal encounter with Newton had far-reaching consequences for 's Gravesande's ideas about natural philosophy. However 's Gravesande certainly knew of Newton's work before this.[16] In May 1714 – a year before he went to England – he had already written a letter to the English scholar in which he humbly offered his services "on all occasions I could be of some use to you in this country".[17]

An occasion for being "of some use" occurred when he was appointed at Leiden University as professor in "mathematics and astronomy" in 1717. It is not completely clear why the board of the university wanted 's Gravesande to take this chair (which had been vacant for years). Without doubt Duyvenvoorde's patronage had been involved; Allamand even maintains that Newton himself had recommended 's Gravesande.[18] In any case, 's Gravesande now was in the position to start spreading Newton's word and already three years later the first part of his famous handbook *Physicis elementa mathematica, experimentis confirmata. Sive introduction ad philosophiam Newtonianam* (1720–1721), appeared; including the description of the demonstration instruments he designed with the instrument maker Jan van Musschenbroek (1687–1748). The book would be reprinted four times and was translated in several languages.

The conclusion that has been drawn from the two trips to England is that two English philosophers, Boyle and Newton, put a decisive stamp on the founding and the development of the Leiden Cabinet of Physics under the direction of de Volder and 's Gravesande. In the remainder we will demonstrate, with the help of some of the instruments of the Cabinet, that also other motives have been involved and that the influence of

15 's Gravesande, *Mathematical Elements of Natural Philosophy, Confirmed by Experiments: Or an Introduction to Sir Isaac Newton's Philosophy... Translated into English by J.T. Desaguliers...*, London, 1720.

16 Maas, *The Man...*, cit. (n. 10).

17 "[...] dans toutes les occasions que je pourai vous estre de quelque utilité dans ce pais"; 's Gravesande, Letter to Isaac Newton, 28 May 1714, in A. Rupert Hall, Laura Tilling, Herbert W. Turnbull (eds.), *The Correspondence of Isaac Newton*, Cambridge, 1959–1977, 7 vols.; v. 6, pp. 144–145; William Burnet, Letter to Johann Bernoulli, 8 April 1714, Ibid., pp. 96–97. See also Richard S. Westfall, *Never at Rest: A Biography of Isaac Newton*, Cambridge, 1980, pp. 741–744.

18 Maas, *The Man...*, cit. (n. 10).

British natural philosophy sometimes worked out in a different manner than has been suggested by historians.

A Showpiece to Attract Students – de Volder's Air-pump

The only source for the foundation of the Physics Theatre – the report of the meeting of the board of the university in which de Volder's request was discussed – does not elaborate on introducing experiments as a means to end dogmatic quarrels. According to the report, de Volder had explained the board that he wanted to show by experiment "the truth and certainty of the postulates and theories presented to the students in Physica theoretica".[19] He was absolutely convinced that the "usefulness and entertainment of the proposed pursuit of experiments" would draw "many students from other universities and schools elsewhere".[20] A main argument, in other words, was simply to attract more students.

In the days of de Volder it was not easy to get hold of physical instruments. There were instrument makers, but they mostly fabricated precious and precise astronomical and mathematical instruments. What de Volder needed was a craftsman who was up to making his experimental classes work. He found that man in Samuel van Musschenbroek, a brass founder in Leiden.[21] Van Musschenbroek made de Volder's first instrument: an air-pump, which would remain in use well into the next century (*Fig. 1*).[22] This air-pump took the centre stage in de Volder's experimental courses. Several pneumatic and hydrostatic experiments were conducted with it, and for instance also dogs and birds were subjected to the vacuum. Using an air-pump as part of the physics curriculum was in the 1670s an innovative idea, though not without precedent. As Wiesenfeldt sets out, de Volder was acquainted with Johann Christoph Sturm (1635–1703) who had introduced experimental courses already at the Nurnberg University Altdorf, including an air-pump, in 1672.[23] The air-pump was in those days

---

19 Governers' Resolutions 3 December 1674, in Philip Christiaan Molhuysen, *Bronnen tot de geschiedenis der Leidse Universiteit*, The Hague, 1913–1927, 7 vols.; v. 3, p. 298.

20 Wiesenfeldt, op. cit. (n. 1), p. 62.

21 About the Van Musschenbroek workshop, Peter de Clercq, *At the Sign of the Oriental Lamp: The Van Musschenbroek Workshop in Leiden, 1660–1750*, Rotterdam, 1997.

22 Peter de Clercq, *The Leiden Cabinet of Physics: A Descriptive Catalogue*, "Museum Boerhaave Communications 271", Leiden, 1997, pp. 14–15.

23 Wiesenfeldt, op. cit. (n. 1), pp. 66–67; Hans Schimank, "Zur Geschichte der Physik in Göttingen", *Rete: Strukturgeschichte der Naturwissenschaften* 3 (1974), pp. 207–252: 208.

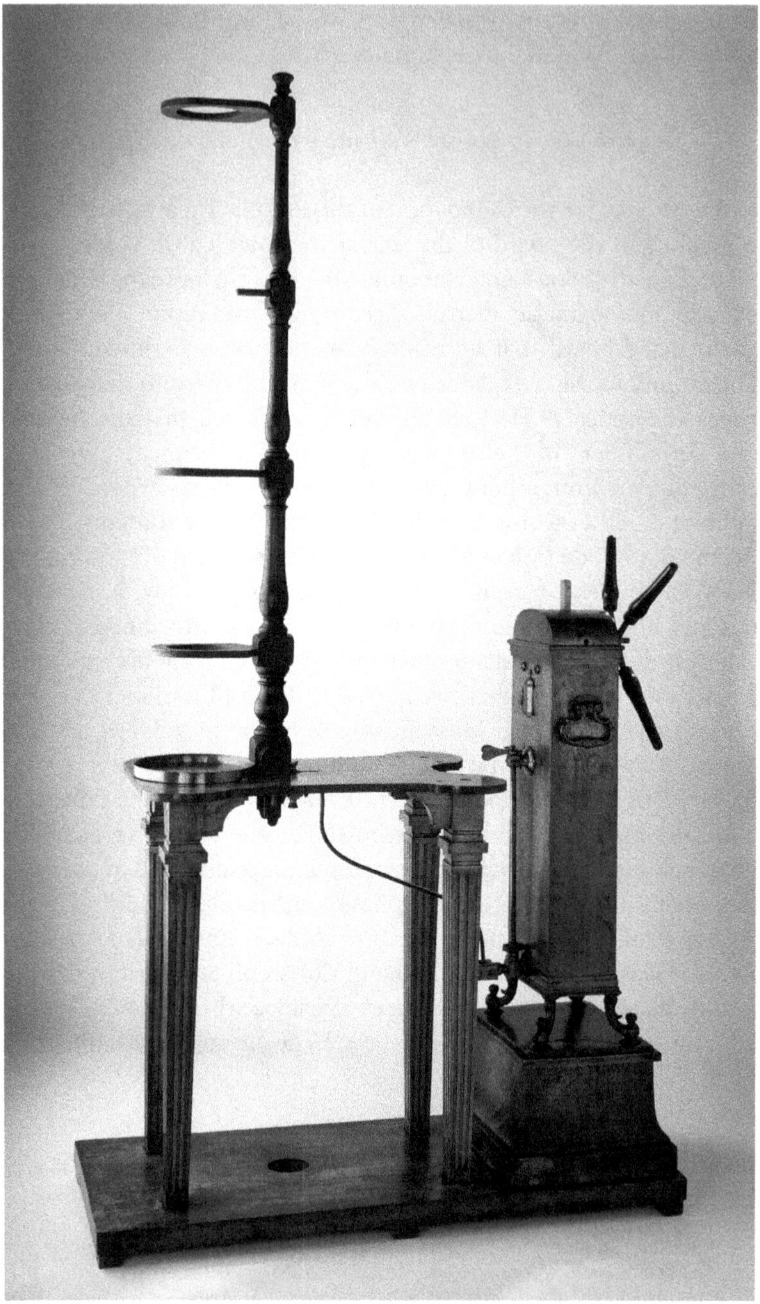

Fig. 1. Single barrel air-pump by Samuel van Musschenbroek, 1675 (Museum Boerhaave, Leiden, inv. no. V09623).

still a scarce and expensive commodity that undoubtedly will have succeeded in drawing the attention of (foreign) students.

Thus the air-pump, which was according to a description from 1711 ostentatiously placed on the raised platform in the middle of the theatre,[24] fitted into the University's policy to collect and display conspicuous collections to distinguish itself from other universities and enhance its reputation as a Dutch – and even European – centre of scholarship and to attract (affluent) students. The anatomical theatre was the first example of practical education.[25] Over the years it gained a reputation, although mostly from its variety of *curiosa*. The botanical garden was enlarged in the years 1687–1688. Built in 1633, the observatory was upgraded in this period – de Volder himself also had instruments bought such as a large sextant.[26] The first call for a chemical laboratory had come up in 1664, and with the installation of the new professor Carel de Maets (1641–1690) the first Dutch university laboratory saw the day of light. Before that there had already been private laboratories in Leiden as the one of de la Boë Sylvius (as we have seen, one of the teachers of de Volder). Leiden University thus tried to present itself as the most exquisite university of Europe.[27] A theatre with a collection of physics demonstration instruments clearly suited this intention. In the Dutch Republic, Groningen followed suit with experimental physics courses in 1697, followed by Harderwijk, Franeker and, finally, Utrecht.

The air-pump remained an important feature of the natural philosophy classes in Leiden. In 1679 Senguerd suggested improvements for the air-pump and had them made by the van Musschenbroek workshop.[28] Later, also 's Gravesande worked out technical advancements and in his turn had them produced by the same manufactory.[29]

---

24 Von Uffenbach, *Merkwürdige Reisen* . . ., cit. (n. 9), pp. 425 and 454.

25 For the history of the anatomical theatre, see Tim Huisman, *The Finger of God*, Leiden, 2009.

26 Wiesenfeldt, op. cit. (n. 1), p. 91.

27 Otterspeer, op. cit. (n. 7), pp. 95–145.

28 Senguerd air-pump is shown for instance in Wolferd Senguerd, *Philosophia naturalis*, Leiden, 1681, p. 65; one of Musschenbroek's pumps after Senguerd's design is in Museum Boerhaave (inv. No. V09551).

29 De Clercq, *At the Sign* . . ., cit. (n. 21).

Magdeburg Hemispheres – The New Philosophy

De Volder was known for his Cartesian sympathies. This adherence has been brought forward as an additional argument to underpin the hypothesis that de Volder introduced the experimental method to get Leiden natural philosophy into calmer waters. After all, in the Cartesian view observations were subordinated to reasoning, which at first glance makes the introduction of experiments no absolute requirement for transferring natural philosophical knowledge to students.

Surprisingly, in his experimental courses de Volder proved to be very critical of Descartes' explanations of natural phenomena. Instead, he closely followed Boyle's experimental program and defended his views (ignoring, however, the religious connotations in Boyle's explanations). De Volder cited in his experimental courses scholars such as Evangelista Torricelli and Blaise Pascal, who just like Descartes and Boyle were spokesmen of the new, mechanistic philosophy and opposed the traditional scholastic views.[30] So, de Volder, rather than being a "Cartesian" in the strict sense proved to be in practice a protagonist of the new philosophy and mechanism in a general sense. Perhaps being a "Cartesian" in those days had a wider meaning than advocating the views of Descartes, just as alleged atheists were often being labeled "Spinozists", irrespective if they really endorsed Baruch Spinoza's views. Descartes had lived for twenty years in the Dutch Republic and was the most well-known (and notorious) representative of the new philosophy. He may, in other words, have functioned as the personification of the new philosophical movement. De Volder's colleague and later successor as director of the Physics Theatre, Wolferd Senguerd, started his career with a carefully built conservative image. This made him the perfect candidate for the Leiden University to counterbalance the progressive de Volder. Senguerd was assigned with "peripatetic physics", though in practice he taught an eclectic mixture of old and new views.[31]

From this point of view, the introduction of the experimental method fitted squarely in de Volder's attempts to spread the new, mechanistic philosophy in general and Boyle's program in particular. In this respect, the influence of British natural philosophy on Leiden physics has been larger

[30] Wiesenfeldt, op. cit. (n. 1), pp. 99–132.
[31] Ibid., pp. 132–157.

than de Volder's qualification as "Cartesian" would suggest. With vacuum experiments conducted for instance with Magdeburg hemispheres (*Fig. 2*), de Volder demonstrated that air possessed "heaviness", and not (peripatetic) "lightness". He showed with a balance that an evacuated hollow sphere weighed less than one filled with air. Air-pressure was explained by de Volder as caused by aether particles in the pores of air, thereby combining Boylean and Cartesian views.[32]

Entertainment in the Theatre of Physics – Magic Lantern

Certainly 's Gravesande, de Volder's de facto successor, regarded demonstrations as a means to make the physics courses more accessible and attractive to wider audiences. In a letter to Newton, he wrote: "as I talk to people who have made very little progress in mathematics I have been obliged to have several machines constructed to convey the force of propositions whose demonstrations they had not understood".[33]

In addition to his more serious demonstrations (see below) 's Gravesande enchanted his students also with "gadgets" like a magic lantern, which projected slides showing images of satyrs, dwarfs and Arcadian landscapes (*Fig. 3*), and anamorphoscopes (where a distorted picture takes on a normal appearance only when seen in an appropriately shaped, mostly cylindrical or pyramidal, mirror).[34] Such demonstrations could hardly have served other ends than to lend 's Gravesande's courses a touch of entertainment and spectacle. It will have brought him extra students and, perhaps not unimportantly, extra income. Incidentally, the inventory-list of the physical theatre of 1705 reveals that also de Volder had disposed of an anamorphoscope-mirror, as well as a magic lantern.[35]

Both 's Gravesande and de Volder created, intentionally or not, dramatic effects with vacuum-pump experiments in which dogs, rabbits, birds and other animals were subjected to the vacuum.[36]

---

32 Ibid., pp. 117, 127.

33 Willem J. 's Gravesande, Letter to Newton, 13/24 June 1718, in A.R. Hall, "Further Newton Correspondence", *Notes and records of the Royal Society of London* 37 (1982–1983), pp. 7–34. The original is in French.

34 De Clercq, *The Leiden Cabinet...*, cit. (n. 22), pp. 108–124.

35 Molhuysen, *Bronnen...*, cit. (n. 19), v. 4, pp. 104–106.

36 Wiesenfeldt, op. cit. (n. 1), pp. 99–132; W.J. 's Gravesande, *Physicis elementa mathematica, experimentis confirmata*, Leiden, 1720–1721, p. 234.

Fig. 2. Magdeburg hemispheres probably by Samuel van Musschenbroek, c. 1675 (Museum Boerhaave, Leiden, inv. no. V09647).

Fig. 3. Projection Lantern by Jan van Musschenbroek, c. 1720 (Museum Boerhaave, Leiden, inv. no. V10916).

's Gravesande the Tinkerer – The Heliostat

's Gravesande, officially appointed professor of astronomy and mathematics in 1717, was only able to use the *Theatrum Physicum* for public lectures after the death of Senguerd in 1724, when he became its director. Already in his early days as a university professor, however, he demonstrated, as mentioned above, physics instruments at home during his lucrative private courses (to attend private courses, students had to pay their professors a fee). More than had been the case in de Volder's or Senguerd's courses, experiments were systematically interwoven with the subjects on the curriculum.

It is clear that the advanced interaction of mathematics and instruments, besides fitting the Newtonian method of investigating nature, expressed perfectly well 's Gravesande's innate interests and gifts. In the above we discussed his fondness for mathematics. Besides that, 's Gravesande loved to work on instruments. Already as a student he had been experimenting with air-pumps. Now as a professor – together with instrument maker Jan van Musschenbroek – he would devise many innovative machines.[37]

It is remarkable how 's Gravesande and van Musschenbroek managed to produce their first series of apparatuses only in three year's time. These were meticulously described by 's Gravesande in his textbook *Physicis elementa mathematica*, which contained numerous detailed engravings of the instruments.

In designing the instruments he cooperated closely with van Musschenbroek, but some instruments definitely reflect his own ingenuity. One of these is also one of the prettiest: the heliostat (*Pl. III*). For an experimenter there was no good alternative for sunlight. However, to get a continuous sunbeam for light experiments proved very difficult. In this heliostat, made not by van Musschenbroek this time but by the clockmaker Jacob van der Cloesen (c. 1690–1766), the mirror is moved by a clockwork in order to follow the sun and give the experimenter a solid beam of light. In designing this elegant, but treacherously sophisticated instrument,

[37] Maas, *The Man*..., cit. (n. 10); de Clercq, *At the Sign*..., cit. (n. 21), pp. 73–102; De Clercq, *The Leiden Cabinet*..., cit. (n. 22); De Clercq, "The 's Gravesande Collection in Museum Boerhaave, Leiden", *Nuncius* 1 (1988), pp. 127–137.

's Gravesande gave an excellent proof of his skills and extensive experience in both instrument designing and mathematical thinking.[38]

's Gravesande and the Pursuit of "True" Knowledge – The Fall Apparatus

A central theme in 's Gravesande's metaphysical writings is his preoccupation with finding true, "unprejudiced" knowledge.[39] Since the interference of the human mind according to him only served to corrupt the study of nature, this in particular should be limited as much as possible. This position led 's Gravesande to two conclusions. Firstly, just like Newton, he rejected deductive reasoning not sustained by observation, in the manner of Descartes. 's Gravesande argued for the outside world, God's creation, as a main source of irrefutable knowledge. With the help of instruments such "irrefutable" knowledge could be produced.

Secondly, 's Gravesande excluded arguments simply based on the authority of a revered scholarly person. He maintained this principle consistently, even if the ideas of Isaac Newton were at stake. 's Gravesande's point of view in the *vis viva* question is striking. In the debate on whether the "force" of an accelerated object increased proportionally with the velocity (*quantitas motus*) or with the square of the velocity (*vis viva*) 's Gravesande – convinced by his own experiments – publicly sided with Leibniz *cum suis*, thereby defying his idol and mentor Isaac Newton. By dropping brass balls of various weights in a layer of clay in the tray of a fall apparatus (*Fig. 4*), 's Gravesande found that the same product of height and weight caused identical impressions in the clay. This could only mean that Leibiniz's *vis viva* (square of the velocity) was correct.

Pious Newtonians like Samuel Clark reacted angrily to 's Gravesande's "treason", especially since far-reaching metaphysical and religious consequences, which they had drawn from Newton's natural philosophy, collapsed if the *quantitas motus* would be untenable.[40] 's Gravesande, however, coolly replied that "real Newtonians don't follow a person but a method: not Leibniz or Newton, but the experiments have the last word".[41]

---

38 De Clercq, *The Leiden Cabinet*..., cit. (n. 22), p. 88; de Clercq, *Het Leids Fysisch Kabinet*, Leiden, 1989; Allan A. Mills, "Portable Heliostats (Solar Illuminators)", *Annals of Science* 43 (1986), pp. 69–406: 383–386.

39 Maas, *The Man*..., cit. (n. 10).

40 Israel, *Enlightenment Contested*..., cit. (n. 3).

41 De Pater, *Willem Jacob*..., cit. (n. 10), p. 19.

Fig. 4. Fall apparatus probably by Jan van Musschenbroek, 1722 (Museum Boerhaave, Leiden, inv. no. V09630).

's Gravesande had published his results in 1722, and from the second edition in 1725 his handbook included the fall apparatus. In 1733 's Gravesande designed yet another instrument to demonstrate the "force" of a moving body. This one was based on the impact made by brass cursors on an iron pendulum-bar in a vertically positioned layer of clay.[42] Thus, not only the *vis viva* became firmly established in 's Gravesande's teaching, but also, implicitly, his position regarding the search for "true" knowledge.

Hydrostatic Balance – A Complete Survey of Mathematical Physics

A well-known fact, but good to emphasize once more, is that 's Gravesande's book was a complete handbook of mathematical physics. Although 's Gravesande is mainly known as propagator of Newton, his exposition contained much more than Newtonian mechanics and optics alone, both in his book and in his demonstrations. The arguably most famous instrument of the Leiden Cabinet of Physics, his hydrostatic balance (nicknamed "'s Gravesande's bucket", *Fig. 5*) had nothing to do with Newtonian physics but was an elegant expression of Archimedes' law. Indeed, quite a large proportion of the instruments 's Gravesande had constructed consisted of tools to demonstrate the working of ancient mechanics (as had been synthesized by early-modern scholars, in particular Guidobaldo del Monte (1545–1607)). Levers, pulleys, wedges and inclined planes expressed the "simple machines" that had its origins in ancient Greek texts.[43] The representation of ancient mechanics is a marked difference with de Volder, who revealed himself in the choice of his experiments predominantly as an advocate of the new philosophy.

's Gravesande will have considered ancient mechanics as inherent to the general survey of mathematical physics that he aimed to offer his students. In this he possibly followed John Keill, who had published books on experimental philosophy that also included the *philosophia mechanica* (as he had named it). Yet, 's Gravesande does not mention Keill in his book. Another motive might have been that by referring to ancient scholarly work, 's Gravesande wanted to appeal to the students who came Leiden University in the first place to be versed in the humanistic culture, the mastery of which belonged to the *savoir vivre* of the higher classes. This

[42] De Clercq, *The Leiden Cabinet...*, cit. (n. 22), p. 43.
[43] Ibid., pp. 23–32.

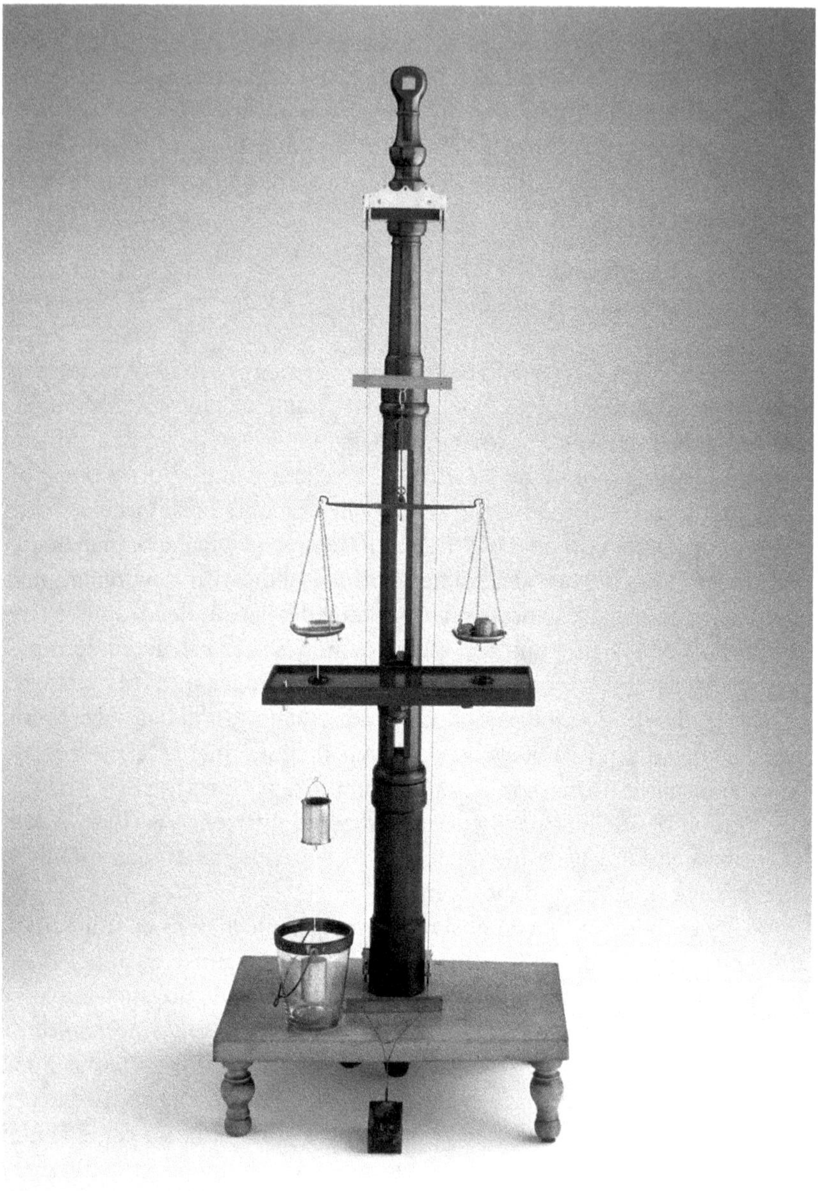

Fig. 5. Hydrostatic balance, probably by Jan van Musschenbroek, 1725–1740 (Museum Boerhaave, Leiden, inv. no. V09295).

could also explain the presence among his instruments of an apparatus to show Hero's fountain.[44] A final reason for the ample attention to the simple machines may have been their practical usefulness. In that case they belonged to the same category as the "useful machines" 's Gravesande added to the Cabinet, which we will discuss next.

Windmill – 's Gravesande's Metaphysics

By the time of Senguerd's death in 1724, 's Gravesande had built up a vast collection of apparatus without parallel in Europe at his own expense for his lessons at home. The board of the University, who wanted "the public lectures to be continued", obviously noticed the successful "experiments in his lectures with large student audiences".[45] Not surprisingly, they appointed 's Gravesande as professor of natural philosophy with responsibility for the theatre. Yet, 's Gravesande was not at all happy with the equipment he found there, as it was more or less the same as de Volder had started with fifty years before.[46] Instead of duplicating the instruments he used at his home, however, 's Gravesande chose to continue the somewhat obsolete demonstrations with air-pumps and only bought a few instruments he did not have already. Actually, he only introduced one really new brand of instruments in the *Theatrum Physicum*: the "useful" machine models that appeared in the theatre from the 1730s onwards. Windmills (*Pl. IV*), a water lifting pump, even a steam-engine, and so on, were apparently meant to have an emblematic character and demonstrated some of 's Gravesande's deep-rooted religious convictions. According to him, God had created an orderly world for mankind to live in. He created nature and its laws in the service of humankind. It was the task of the philosopher to foster the well-being and happiness of others and to search for ways to take optimal profit from nature.[47] The models, then, showed how man might benefit from nature and the natural laws in the pursuit of his own well-being, a theme that cannot be considered typically "Newtonian".

Indeed, as Tiemen Cocquyt elaborates in this volume, 's Gravesande matched actions to words, as regards the obligations of the philosopher,

---

[44] Ibid., p. 82.
[45] Molhuysen, *Bronnen*..., cit. (n. 19), v. 1, p. 30.
[46] De Clercq, *The Leiden Cabinet*..., cit. (n. 22), pp. 6 and 7.
[47] De Pater, *Willem Jacob*..., cit. (n. 10), pp. 26–30.

for he was active as technical advisor for the States of Holland in matters of water management. Cocquyt suggests that some of the machine models might have been research or patent models, rather than merely educational tools, reflecting 's Gravesande's hydraulic involvements.[48]

Conclusion

In this article we have used some of the eye-catching instruments of the Leiden Cabinet of Physics to "reveal" possible motives of de Volder and 's Gravesande, which put a considerable mark on the Cabinet's development. With the help of the instruments, we were able to argue that de Volder perhaps should be considered as a representative of the "new philosophy", rather than a Cartesian in the strict sense, and have nuanced the "Newtonian" character of 's Gravesande's instruments. 's Gravesande obviously not only wanted to teach Newtonian views, but intended to give a complete physics survey. Some of his instruments even expressed convictions that defied those of Newton. We argued that the collection of machine models expressed 's Gravesande's religiously buttressed opinion that it was man's task to take optimal profit from the natural world to enhance his well-being.

We concluded additionally that, besides high-principled scientific and metaphysical motives, more earthly motives determined the fate of the Leiden Cabinet of Physics as well – and maybe even to a stronger degree. Firstly, the continuous improvement of the experimental courses naturally gave 's Gravesande also a pretext for pursuing his own passion for designing instruments. Next, we argued that the Cabinet – maybe even above all – was a tool that anticipated the wishes and needs of the students. The University functioned not merely as a disinterested educational institute. It was also an enterprise that wanted to attract as many (affluent) students as possible, and its professors were entrepreneurs whose income partly depended on the number of students who attended their courses. Just like ordinary human beings, the Leiden physicists seem to have been driven by passion, money, fame and a successful career as much as by high-standing principles. The *Theatrum Physicum* was their enterprise in experiments.

[48] See *Infra*, pp. 79 ff.

We do not pretend to have given a complete survey of all purposes for which the Cabinet has been used in the times of de Volder and 's Gravesande. Yet what we have argued in conclusion is that the Leiden Cabinet of Physics was not simply a linear transformation of British natural philosophical ideas across the Channel. The Leiden Cabinet of Physics was enormously successful and its instruments were copied all over Europe. Undoubtedly, a close study of other collections will reveal that these also were moulded by very particular local needs and preferences, and by entrepreneurs who perhaps, behind a façade of grave philosophical and religious arguments, satisfied their hunger for fame and money.

NEW LIGHT ON THE CABINET OF PHYSICS OF PADUA

Sofia Talas

"The experimental physics room was established a few years ago by the Marquis Poleni; he has gathered a large collection of machines of all sorts, made in France, in England, in Holland, under the eye of the best Physicists; many of them were devised or improved by Poleni himself, and I do not know of a more beautiful cabinet of Physics".[1] With these words, the French astronomer Joseph-Jérôme de Lalande described the Cabinet of Physics he visited and admired at the University of Padua in 1765. The collection had been set up for the lecture courses on experimental natural philosophy that had been introduced at the University of Padua a few years before, with the creation of a chair of Experimental Philosophy in 1738. Giovanni Poleni, who was assigned the new chair, gathered a first group of machines within a couple of years and he continued to enrich the collection until his death, in 1761.

Several researches have already focused on Giovanni Poleni not only as a scientist, but as a philologist and an acknowledged expert in architecture and classical antiquity as well (*Fig. 1*).[2] Let us only remember that Poleni was born in Venice in 1683 and that he published on a wide variety of scientific topics ranging from calculating machines to navigation instruments, from sundials to thermometers, from astronomy to hydraulics. A fellow of the Royal Society and member of the Academies of Sciences of Paris, Berlin and Saint Petersburg, he was in touch with many prominent figures of his time, such as Leonhard Euler, Pierre-Louis

---

[1] "La salle de physique expérimentale, fut établie il y a quelques années par le Marquis Poleni; il y a rassemblé une ample collection de machines de toute espece, faites en France, en Angleterre, en Hollande, sous les yeux des meilleurs Physiciens; plusieurs ont été imaginées ou perfectionnées par M. Poleni lui-même, & je ne connois gueres de plus beau cabinet de Physique"; Joseph-Jérôme de Lalande, *Voyage d'un François en Italie, Fait dans les années 1765 et 1766*, Paris, 1769, 8 vols.; v. 8, pp. 285–286.

[2] Giuseppe Grioli, Augusto Ghetti *et Alii, Giovanni Poleni (1683–1761) nel bicentenario della morte*, Padua, 1963; Maria Laura Soppelsa (ed.), *Giovanni Poleni, idraulico, matematico, architetto, filologo*, Padua, 1988; Gian Antonio Salandin, Maria Pancino, *Il Teatro di Filosofia Sperimentale di Giovanni Poleni*, Trieste, 1987; Gian Antonio Salandin, Sofia Talas, "Giovanni Poleni", in *La curiosità e l'ingegno* (exhibition catalogue), Padua, 2000, pp. 85–91; G.A. Salandin, S. Talas, "Strumenti e macchine", Ibid., pp. 223–243.

Fig. 1. Portrait of Giovanni Poleni, in Bartolomeo Gamba, *Galleria dei letterati ed artisti illustri delle provincie veneziane nel secolo decimottavo*, Venezia, 1824, v. 2, p. 99 (Venice, Biblioteca Nazionale Marciana).

Moreau de Maupertuis, Jean Antoine Nollet, Nicolaus I Bernoulli, Jean Jallabert, Roger Boscovich and Pieter van Musschenbroek. He was successively assigned several chairs at the University of Padua: the chair of *Astronomia e meteore* in 1709, the chair of *Philosophia ordinaria in secundo loco* in 1715 and, in 1719, the prestigious chair of mathematics, that had been occupied by Galileo and other outstanding scientists, like Jacob Hermann and Nicolaus I Bernoulli.

It is worth pointing out that, at the beginning of the eighteenth century, the course of *Philosophia ordinaria in secundo loco* comprised elements of natural philosophy. According to traditional standards, the teaching was based on the reading of Aristotle's works. No experiment-based course existed. There was however a growing need for a change. Experimental natural philosophy had been practised in Venice from the late seventeenth century, in particular at the Sarotti Academy, founded in 1681 on the model of the Royal Society.[3] This means that the Venetian patricians, who at that time were in charge of the University of Padua, were aware of the development of the new scientific practice. From 1715 onwards, several documents were proposed, stressing that it was urgent to achieve a general reform of the University in order to eliminate old-fashioned chairs and create new ones.[4] Two of the most important of these proposals, written respectively by Scipione Maffei in 1715 and by Giovanni Francesco Pivati in 1738, both suggested the creation of a chair of experimental philosophy. Such a chair was finally institutionalized on 27 November 1738 and, a couple of months later, on 12 February 1739, it was assigned to Poleni.

New researches, mainly based on Poleni's unpublished correspondence, have been carried out on the Venetian scientist, aiming at analysing in detail the creation of the university Cabinet of Physics in Padua. The first results of these researches, which are still in progress, are presented in this paper. We will describe at first the role of the Venetian authorities, who were very concerned with making the new chair of experimental philosophy as successful as possible. We will then outline the influence of the European context upon Poleni's lecturing on experimental philosophy. Finally, by examining the contacts Poleni set up in Italy and abroad, we

[3] The meetings of this academy were attended for two years by Denis Papin. See for instance Denis Papin, *Continuation du Digesteur ou manière d'amolir les os*, Amsterdam, 1788, section III, pp. 164–240, describing the "Expériences faites dans l'Académie de M. Sarotti de Venise".

[4] Piero Del Negro, "L'Università", in Girolamo Arnaldi, Manlio Pastore Stocchi (eds.), *Storia della cultura veneta*, Vicenza, 1976–1986, 10 vols.; v. 5/I "Il Settecento", pp. 47–76.

will see how Poleni acquired, little by little, the instruments he needed for his lectures and his researches.

The Role of the Venetian Republic

Dozens of letters attest the serious involvement of the Venetian republic officials in all the steps that led to the creation of the Cabinet of Physics at the University of Padua. In November 1738, already, as the new chair was just about to be institutionalised, Giovanni Francesco Morosini, one of the *Riformatori allo Studio di Padova*, the elite group of three Venetian patricians who ruled the University of Padua, personally wrote to Poleni to ask him about "a valuable person to give the course of Experimental Philosophy with honour".[5] Poleni answered by providing the names of Christian Wolff, John T. Desaguliers, Willem 's Gravesande and Pieter van Musschenbroek who, according to him "perform experiments as an art".[6] In this letter, Poleni went on to give more details about the qualities required for such a task: the new professor should have experience of practicing with instruments and "ingenuity in his hands",[7] in order to be able to use the instruments fruitfully, as well as a good theoretical knowledge.

Poleni himself met all these requirements, as he had been successfully practicing experimental natural philosophy with his own instruments for years, from the early eighteenth century, when he was still living in Venice. Quite soon, the Venetian authorities thus turned to regard him as a possible candidate. But they did not make up their mind without submitting him to a close examination. Pivati, the promoter of the 1738 reform that led to the creation of the chair of experimental philosophy, was charged with this task. He lived in Venice and he was a kind of "minister" of the *Riformatori*. Very interested in experimental philosophy, Pivati was to announce, in 1747, that he had infused medicines through electrical discharges.[8] It is well known that his claim was to be demolished by

[5] "[...] soggetto valevole a sostenere onorevolmente la lezione di Filosofia Sperimentale"; Giovanni Francesco Morosini, Letter to Poleni, 13 November 1738: Venice, Biblioteca Marciana (BMVe), ms. it., IV, 592 (= 5555), f. 229.

[6] "[...] questi degli esperimenti fanno un'arte"; Giovanni Poleni, Letter to Morosini, undated: Ibid., f. 191.

[7] "[...] ingegno alle mani"; Ibid.

[8] Giovanni Francesco Pivati, *Lettera intorno all'Elettricità medica*, Lucca, 1747; G.F. Pivati, *Riflessioni fisiche sopra la medicina Elettrica*, Venice, 1749. He also wrote a *Nuovo dizionario scientifico e curioso sacro- profano* [Scientific and curious sacred and profane dictionary], Venice 1746–1751, 10 vols.

Jean Antoine Nollet, when he travelled through Italy in 1749.[9] But let us go back to the 1730s. It seems that Pivati "examined" Poleni so closely that he hurt him. He thus wrote to Poleni in December 1738:

> You tell me that I observed you in Padua with a microscope. It must have been a bad one because it did not enlarge the object at all. And if I now observe you with a telescope, my lack of practice makes that, in looking at you, I have put the objective near the eye and the ocular pointing to the object, so that instead of seeing you as you are, I have seen you inferior to your great merit, but this happened because my eye is not able to see the most admired objects of the world like you.[10]

When Poleni was officially assigned the new chair, Pivati was still very active as one of the links between Poleni and the Venetian authorities, in particular as regards the requests that Poleni insisted upon in the numerous letters he sent around, not only to Pivati, but to other Venetian officials and to personal friends as well.

In those letters, Poleni explained his three major worries related to the new chair. First of all, he needed someone who could help him "when we will prepare the machines, when we will carry out the experiments and when we will try them privately before presenting them to the public".[11] As he was also asked to continue his lectures on mathematics, Poleni complained that "Old as I am, and with these health problems [. . .] I did not feel strong enough to accept such a burden [. . .] I shall say as the English say, let's see".[12] Poleni also insisted on getting an adequate place to hold the new lectures, refusing to do it at home, as he regarded it as

---

[9] Jean Antoine Nollet, *Essai sur l'électricité des corps*, 2nd ed., Paris, 1750. See also John. L. Heilbron, *Electricity in the 17th and 18th Centuries*, New York, 1999, and Paola Bertucci, *Viaggio nel paese delle meraviglie. Scienza e curiosità nell'Italia del Settecento*, Turin, 2007.

[10] "Ella poi mi dice che in Padova l'ho riguardata col microscopio. Bisogna che fosse molto poco buono perché di nulla m'ingrandiva l'oggetto. E se qui la osservo col telescopio, la mia poca pratica fa che nel riguardare Vostra Signoria Illustrissima, io abbia posto l'obiettivo all'occhio e l'oculare verso l'oggetto, quindi è che invece di rappresentarmi la sua degnissima persona tal qual si è, me l'ha rappresentata di gran lunga inferiore al suo gran merito, ma questo forse sarà avvenuto per mio difetto, che l'occhio mio manchevole non può, né sa quanto basta distinguere gli oggetti del mondo tutti ammirati, com'è Vostra Signoria Illustrissima"; Pivati, Letter to Poleni, 29 December 1738: BMVe, ms. it., IV, 592 (= 5555), f. 236.

[11] "[. . .] quando apparecchieremo le machine e quando dirigeremo gli esperimenti e quando li proveremo privatamente fra noi avanti di mostrarli al pubblico"; Poleni, Letter to Gadaldini, 18 marzo 1739: BMVe, ms. it., IV, 592 (= 5555), ff. 185–186. Agostino Gadaldini was the secretary of the *Riformatori*.

[12] "Io veramente in questa età, e con questi acciacchi [. . .] non mi sentiva in istato di assumere questo peso [. . .] Dirò come dicono gli Inglesi, vedremo"; Poleni, Letter to Maffei, 23 February 1739: BMVe, ms. it., X, f. 94.

uncomfortable and unfair for students. He thus started lecturing at the University, but described "the sorrow of those many who were there to watch, but could not see anything".[13] Finally, Poleni asked for funds, as he needed instruments for teaching. He already had some instruments of his own, but he bitterly complained that these were hardly sufficient for even a few lectures. The *Riformatori*, before Poleni had been assigned the chair, had already got in touch with the heirs of Cristino Martinelli, a Venetian patrician and member of the Sarotti Academy, in order to buy some of his instruments. Poleni was asked to examine these devices and he chose five of them, i.e. a double-barrelled air-pump, very similar to Hauksbee's design, a mirror, a compass made by Richard Whitehead, an artificial magnet and a camera obscura (*Fig. 2*). "The Martinelli machines", according to Poleni, "are beautiful, but their use can hardly fulfil one lecture", so that his students, he said, were full of "pity" when they attended his lessons.[14]

Here again, the serious interest of the Venetian government in the new experimental philosophy lectures emerges quite clearly: in a very short time, within a few months, the *Riformatori* fulfilled all Poleni's requests. The scholar was provided with assistants he personally selected. One of them, Giovanni Antonio Dalla Bella, was to lecture on experimental philosophy in Lisbon from 1766, setting up a rich cabinet of physics that was moved to Coimbra in 1772. Poleni was also assigned funds to buy instruments on a regular basis. The university Cabinet of Physics in Padua was thus to be one of the first in Europe to be fully financed by public funds. As for the lecturing place, Poleni was charged, together with his friend, the physician Giovanni Battista Morgagni, with planning and supervising renovation works in the historical building of the University, the *Palazzo del Bo*. Specially designed rooms were set up to store the instruments and carry out experiments – a dark room for instance was devoted to optical experiments – and a theatre was built for the lectures. In the very detailed letters that Poleni sent to Venice to describe how the works were going on, the Venetian scholar also explained that he limited the theatre to one hundred seats, because "if the place [the theatre] was larger, part of the

[13] "[...] il dispiacere di que' molti, li quali stavano per vedere, né però vedevano punto"; Poleni, Letter to Bianchi, 1 July 1739: Venice, Archivio di Stato (ASVe), *Riformatori allo Studio di Padova*, file 211. Agostino Bianchi succeeded Gadaldini as secretary of the *Riformatori*.

[14] "Le macchine Martinelli [...] sono belle, ma il loro uso [...] appena può esser materia di una lezione"; Ibid.

Fig. 2. Double-barrelled air-pump bought by the Venetian *Riformatori* from the heirs of Cristino Martinelli (Museum of the History of Physics, University of Padua).

audience would be too far and could not learn, as the experiments are meant to be seen".[15]

Poleni's Experimental Philosophy Lectures in the European Context

Very concerned about the place for the lectures and the instruments to be used, Poleni appears to have been very keen to make his teaching on experimental philosophy as useful as possible for the students. His way of lecturing on experimental natural philosophy mainly followed the Anglo-Dutch Newtonians. Desaguliers but, above all, 's Gravesande and Musschenbroek are mentioned often and enthusiastically in his letters. Poleni wrote for instance in 1739 that "Experiments cannot be performed more than once a week, and *il signor* 's Gravesande doesn't perform them more often".[16] Moreover, the manuscript of a textbook, never completed, where Poleni gave detailed descriptions of his weekly lectures, was titled *Physices elementa mathematica, experimentis confirmata*, just like 's Gravesande well known treatise.[17] The lectures themselves had the typical structure of the experimental philosophy lectures of those years, with definitions, descriptions of experiments and instruments, and explanations. His textbook gave a full bibliography of the books he based his lectures on and, here again, the names of 's Gravesande, Musschenbroek and Desaguliers recur very often, together with other scientists like Nollet, Rohault, Newton or Kepler.

Discussing how important it was to have good and precise instruments, Poleni pointed out how expensive Musschenbroek's, 's Gravesande's and Desaguliers' machines were.[18] He underlined that "the machines we need to have made for the beginning (and I say for the beginning) are those for mechanics [...] that is, a part of those the well-known 's Gravesand also

---

[15] "[...] se il luogo fosse maggiore, quelli che fossero di più per necessità dovrebbero essere più lontani, né sarebbero a portata di ricavar frutto, perché gl'esperimenti sono per quelli che stanno vicini a poterli vedere"; Giovanni Battista Morgagni, Giovanni Poleni, Letter to the *Riformatori*, 13 September 1739: ASVe, Riformatori allo Studio di Padova, file 211.

[16] "[...] non si ponno fare gli esperimenti se non una volta per settimana, né più di una volta per settimana li fa il signor 's Gravesande a Leida"; Poleni, Letter to Zendrini, 7 February 1739: BMVe, ms. it., IV, 643 (= 5504), f. 14. Bernardino Zendrini was the "public mathematician" of the Republic of Venice, officially responsible for the "waters" of the region.

[17] Poleni, "Physices elementa mathematica, experimentis confirmata": BMVe, ms. lat., cl. VIII, 144–146 (= 2721–2723).

[18] Poleni, Letter to Zendrini, 12 February 1739: BMVe, ms. it, IV, 643 (= 5504), ff. 19–20.

starts with".[19] Furthermore, many of Poleni's instruments that were made by local craftsmen are based on Dutch models. His central forces machine and the apparatus for studying the elasticity of materials, for instance, were built according to 's Gravesande's design, while the *macchina divulsoria*, the apparatus that Poleni used to study the strength of materials, when he worked on the restoration of Saint Peter's Cupola in Rome, was very similar to Musschenbroek's original model (*Fig. 3*).[20]

Very close on the one hand to the English and Dutch lecturers, Poleni did not share, on the other, all the views that were quite common in those years. Of course, according to the fashion of the time, he carried out spectacular experiments to catch the attention of his public – he used for instance a fountain of Hero, pouring in water and observing how a jet of wine was spouted out –, thus combining "utility with pleasure", as Nollet

Fig. 3. Machine for studying the elasticity of chords, based on 's Gravesande design (Museum of the History of Physics, University of Padua).

[19] "[...] si dovrebbero far fare per il principio (dico per il principio) le machine meccaniche [...], cioè parte di quelle, con cui principia anco il celebre Gravesand"; Poleni, Letter to the *Riformatori*, undated: ASVe, Riformatori allo Studio di Padova, f. 97.

[20] Poleni, *Memorie istoriche della gran cupola del Tempio Vaticano*, Padua, 1748, pp. 82–90.

suggested.[21] However, while some scientists, like Nollet himself, proudly opened their experimental philosophy lectures to everyone, in particular to women, Poleni stressed

> the great difficulty in practising Experimental Philosophy. Practising it (as an example) as some do, at home, with the participation also of women, to whom nothing difficult has to be presented, with the doors closed, with a certain number of machines and using his own native tongue, is one thing. But it is much more difficult and burdensome to practise Experimental Philosophy in all its length: with dignity and knowledge, and also for those who have already studied and already know about Physics and Mathematics. And this is to be achieved in a Theatre open to all, in the public light and in such a constitution.[22]

Poleni's Acquisitions for the University Cabinet of Physics

Insisting upon the quality and "dignity" of his lecturing on experimental philosophy, Poleni of course needed numerous and good-quality instruments. Moreover, as he was a scientist and a researcher himself,[23] he was also interested in having instruments for his own researches. His correspondence, here again, helps us to understand how, little by little, Poleni set up the University of Padua Cabinet of Physics.

Let us say a few words, at first, about those of Poleni's instruments that were built by local craftsmen, who were not professional instrument-makers. These instruments were based, as we said, on the main physics treatises of those years, though Poleni also designed a few devices himself.[24]

---

[21] J.A. Nollet, *Programme ou idée générale d'un cours de physique expérimentale*, Paris, 1738, p. xi.

[22] "[...] grave difficoltà dell'esercitare la Sperimentale Filosofia. Il farla (per cagion d'esempio), come alcuni nella sua casa privata coll'intervento anche di donne, cui niente di difficile deve essere presentato, a porte chiuse, con un certo numero di macchine ed adoperando la sua nativa lingua, è un fatto. Ma un fatto di gran lunga più gravoso e difficile si è l'esercitar la Sperimentale Filosofia secondo tutta la sua estesa: con dignità, con dottrina ed a portata anche di quelli, che hanno già studiato per rendersi intelligenti di Fisica e di Matematica. E tale incontro ben andiviene in un Teatro a tutti aperto in questa pubblica luce ed in questa costituzione"; Poleni, Letter to the *Riformatori*, May 1761: BMVe, ms. it., X, 313, f. 78.

[23] This was not always the case, as the professors of experimental physics were not supposed at that time to carry out researches at all; Heilbron, op. cit. (n. 9), pp. 134–137.

[24] For instance, Poleni devised an apparatus for studying the velocity of falling bodies, where he used a kind of mercury chronometer for measuring time (see Salvatore dal Negro, *Nuovo oligocronometro*, Padova, 1809), and a device to draw logarithmic and tractorial curves; Poleni, *Ad virum celeberrimum Jacobum Hermannum. Epistola, in qua agitur de organica curvarum tractoriae, atque logaritmicae constructione. Accedunt Problematum*

Several different craftsmen usually contributed to the construction of a single instrument. To make his central forces machine, for instance, Poleni had a goldsmith, a blacksmith, a wood-turner and a joiner working for him.[25] The quality of the surviving instruments suggests that Poleni must have been skilful enough successfully to supervise and guide their work. He was indeed at that time regarded as an authority in this field, renowned throughout Europe. His correspondence with Giovanni Giacomo Marinoni, mathematician, topographer and astronomer at the Imperial Court in Vienna, is paradigmatic in this sense. In 1725, for instance, Marinoni asked Poleni's advice for building a double-barrelled air-pump on the model of Hauksbee's, focusing on very technical details that show the high level of knowledge Poleni was supposed to have.[26] Moreover, in 1738–1739, Marinoni and Poleni thoroughly discussed catadioptric and dioptric telescopes made by different makers. In particular, Marinoni asked Poleni to examine personally Domenico Selva's instruments in Venice,[27] as he was "persuaded that mister Scarlett of London, the one in Scotland, and *domino* Salvi [Selva] of Venice have secrets unknown to me, that they keep with great jealousy".[28]

The locally-made instruments compose quite a large part of Poleni's Cabinet of Physics – about 300 of the 386 natural philosophy instruments listed in his catalogue from 1741 to 1761 –, but Poleni also bought instruments abroad. We will not discuss here his purchasing twenty-four instruments from the Musschenbroek workshop in Leiden and six devices from Nollet (*Pl. V*), as this has already been studied in the past years,[29] but rather focus on the instruments Poleni bought in Italy. As a matter of fact,

atque Theorematum de Curva Tractoria, a Celeberrimus Geometris propositorum, demonstrationes, Padova, 1728.

25 "Note e ricevute delle spese fatte per l'artificiosissima machina, con cui si fanno e si mostrano gli esperimenti delle forze centrali": Padua, Archivio Università di Padova (AUP), busta 591.

26 Giovanni Giacomo Marinoni, Letter to Poleni, 24 March 1725: Verona, Biblioteca Civica (BVCr), Carteggio Poleni, v. 8; Marinoni, Letter to Poleni, 28 April 1725: BMVe, Cod. It., IV, 642 (= 5503), f. 81; Marinoni, Letter to Poleni, 23 June 1725: BCVr, Carteggio Poleni, v. 8.

27 Marinoni, Letter to Poleni, 21 March 1739: BMVe, ms. it, X, 324 (= 6666), ff. 165, 166, 167, 169, 170.

28 "[...] persuaso che mister Scarlett di Londra, quello di Scozia, ed qui il domino Salvi di Venezia vi abbian arcani a me ignoti, che da essi si sopprimono con somma gelosia"; Marinoni, Letter to Poleni, 2 May 1739: BMVe, ms. it, X, 324 (= 6666), f. 167. The maker in Scotland was probably James Short.

29 See note 2 and Peter de Clercq, *At the sign of the Oriental Lamp. The Musschenbroek workshop in Leiden, 1660–1750*, Rotterdam, 1997.

no workshop in Italy, in those years, could provide the whole range of instruments needed for a complete course on natural philosophy, so that Poleni acquired instruments from the most disparate people.

In 1742, for instance, he wrote to Antonio Rossi, a student of his who had moved to Turin,[30] asking about the situation of experimental philosophy there, "where I know it is flourishing. I would also like to know whether there are skilled craftsmen to make machines for this science".[31] He then bought from Giulio Accetta, professor of mathematics at the University of Turin, a crane and other models of machines made by Accetta himself.[32] Bartolomeo Ferracina, mechanician and clock-maker, who restored the tower clock of Piazza San Marco in Venice in the 1750s, also made a few instruments for Poleni: a dilatometer for instance, based on Musschenbroek's design but including Poleni's own improvements,[33] and the model of a pile-driver that Ferracina invented and used when he worked on the reconstruction of a bridge designed by Palladio in Bassano (*Pl. VI*). Two other models of pile-drivers were sent to Poleni by Count Carlo Gazzola, a Veronese nobleman involved in the socially-useful applications of experimental philosophy.[34] From the Duchy of Modena, Poleni received a balance made by an artillery colonel, Domenico Corradi d'Austria. Poleni also entrusted to a Venetian glass factory, Briatti, the making of a couple of crystal-ware containers for experiments on optics and hydrostatics. In Parma, Jacopo Belgrado *s.j.*, professor of mathematics at the local university, was enthusiastically interested in experimental philosophy and wrote to Poleni in 1742, "I try through public activities and some private experiments to stimulate the good taste in a city, where I certainly could not find it".[35] He made a sophisticated model of a fire-machine for Poleni in 1744. Belgrado discussed instruments and makers at length with Poleni, mentioning for instance Pietro Casati, and specifying that he had bought

30 Poleni, Letter to Gadaldini, 18 March 1739: BMVe, ms. it., IV, 592 (= 5555), ff. 185–186.

31 "[...] dove so che tal studio è in fiore. Saprò anche volentieri se vi sieno ora costì valenti artefici per far machine ad uso di quella scienza"; Poleni, Letter to Rossi, 13 January 1742: BCVr, Carteggio Poleni, v. 3.

32 Poleni, "Conti di spese per macchine": AUP, busta 591, and Poleni, "Indice delle macchine": BMVe, ms. it., III, 54–55 = 4969–4970, cl. IV, 626 = 5488, 636 = 5497.

33 Poleni, *Memorie istoriche...*, cit. (n. 20), pp. 55–62.

34 Carlo Gazzola, Letter to Poleni, 18 August 1748: BCVr, busta 217. My warmest thanks to Ivano Dal Prete who sent me a copy of Gazzola's letter and a drawing, thus helping the identification of one of Poleni's pile-drivers with one of those sent by Gazzola.

35 "Io m'ingegno con pubbliche funzioni, e con qualche privato sperimento d'eccitare il buon gusto in una città, in cui certo non ho saputo trovarlo"; Jacopo Belgrado, Letter to Poleni, 7 August 1742: BCVr, Carteggio Poleni, v. 7.

some barometers from his workshop in Como. Poleni answered that he knew Casati very well, regarding him as extremely skilful.[36] The letters the two men exchanged show quite clearly that Belgrado respectfully regarded the Venetian scholar as his master. Quite pragmatic, Poleni also used for teaching some purchases he regarded as relics, such as a compass made by "the famous Venetian Bernardo Facini", a lens of "28 palms, worked by the Illustrious Geminiano Montanari" (*Fig. 4*), and a telescope signed by Giuseppe Campani.[37]

Of course, Poleni got in touch with the very few specialised makers active in the country as well. One of these, Domenico Selva – whose skilfulness,

Fig. 4. Lens signed "Geminianius Montanarius I.V.D. Mat.e Profess. Bon.e 1676", bought by Poleni from the Paduan nobleman Giovanni Lorenzo Orsato in the mid-1740s (Museum of the History of Physics, University of Padua).

[36] Poleni, Letter to Belgrado, 11 August 1742: BCVr, Carteggio Poleni, v. 7.

[37] Poleni, "Indice delle macchine", cit. (n. 32).

as we have seen, was discussed by Poleni and Marinoni in their letters – was an optical maker in Venice with an excellent reputation. From this workshop, Poleni bought some mirrors, a magic lantern, a prismatic lens and a telescope. As for thermometers, barometers and other glass instruments, Poleni acquired many of them from the Pietro Casati mentioned by Belgrado, but also from Girolamo Castelnuovo and the brothers Galli, all active in Como, which was to become one of the main European centres for this kind of craftsmanship (*Fig. 5*). Poleni bought some other instruments from Carlo Antonio Castelnuovo, who had moved from Como to Milan.[38] Castelnuovo, who held public demonstrations on electricity, did not limit his production to glassware and proposed a wider range of instruments, providing Poleni with a solar microscope, a prism, an artificial magnet and some accessories for electrical experiments.

According to the researches carried out as of now, it seems that Poleni did not find other Italian professional instrument-makers but, through Antonio Agelli, a member of the Inquisition who had moved from Venice to Florence, he got in touch with a particularly skilful and talented maker, Philippe Vayringe.[39] Born in 1684 in Lorraine, into a family of peasants, Vayringe had worked as a locksmith and a clock-maker.[40] Sent by Leopold, Duke of Lorraine, to England, in order to learn instrument-making, he worked at Desaguliers' workshop. According to his autobiography, as he arrived in London in 1721:

> Desaguliers hosted me. He taught me Geometry and the principles of Algebra, and methodically taught me the different uses of all the machines he used in the two courses of experimental physics he gave every year. But what was mostly advantageous for me was that this skilled Professor ordered all his workers to make, under my direction, a set of machines similar to the one that formed his Cabinet. As most of these were composed, I found the way to simplify them and to make them even more efficient than before. After thirteen months of work, I was ordered to go back to Lorraine.[41]

---

[38] Ibid.

[39] BCVr, Carteggio Poleni, v. 3; BMVe, ms. it., IV, 643 (= 5504).

[40] An autobiography of Vayringe was communicated by his friend Valentin Jamerey-Duval to Augustin Calmet, who included it in Augustin Calmet, *Bibliothèque lorraine*, Nancy, 1751, pp. 987–999.

[41] "Desaguliers me reçut chez lui en qualité de Pensionnaire. Il m'enseigna la Géométrie & les principes de l'Algèbre, & m'apprit méthodiquement les divers usages de toutes les machines, dont il se servoit dans les deux cours de Physique expérimentale, qu'il donnoit chaque année. Mais ce qu'il y eut de plus avantageux pour moi, c'est que cet habile Professeur ordonna à tous ces ouvriers de travailler, sous ma direction, à construire un assortiment de machines, égal à celui qui formoit son Cabinet. Comme la plupart étaient

Fig. 5. Tantalus' vases by the brothers Galli in Como. These vases were bought by Poleni for the University Cabinet of Physics in 1744 (Museum of the History of Physics, University of Padua).

With such a training, Vayringe made for the Duke Leopold, at the castle of Lunéville, a complete set of instruments in addition to those he brought from London, thus arousing the admiration of Voltaire, who wrote for instance in 1735:

> I was at the court of Lorraine [...]. There is an admirable place established for sciences, not well known and even less used. It is a large room all furnished with the new physics experiments, and in particular with everything that confirms the Newtonian system. There are about ten thousand crowns worth machines of all kind. A simple locksmith become philosopher, and sent to England by the late Duke Leopold, has made most of these machines, and demonstrates with them very neatly. There is nothing similar to this establishment in France.[42]

composées, je trouvai le moyen de les simplifier, & de rendre même leurs effets plus efficaces qu'auparavant. Après treize mois de travail, j'eus ordre de retourner en Lorraine"; Ibid., p. 992.

[42] "J'ai été à la cour de Lorraine [...]. Il y a là un établissement admirable pour les sciences, peu connu et encore moins cultivé. C'est une grande salle toute meublée des expériences nouvelles de physique, et particulièrement de tout ce qui confirme le système

Not only a skilled instrument-maker but a "philosopher" as well, Vayringe also gave lectures on natural philosophy at the Académie de Lunéville from 1729, when the Academy founded by Duke Leopold was reorganised by his successor, Francis III.

When the Lorraine Court moved from Lunéville to Florence in 1737, the collection of instruments was transferred to Palazzo Pitti and, though invited to stay in Lunéville by the new lord, Stanislas of Poland, Vayringe decided to move to Florence too. During the trip, he stopped in the Netherlands and met 's Gravesande in Leiden and Musschenbroek in Utrecht. Very active, he started offering in Florence the lectures he had held in Lunéville, publishing in 1739 a bilingual French and Italian version of the syllabus of his course with the title *Cours de Philosophie mécanique et expérimentale – Corso di filosofia meccanica ed esperimentale*. This was on the model of Desaguliers' English and French catalogue of experiments, published in London in 1713 with exactly the same title, *Course of mechanical and experimental philosophy*. In Florence, however, Vayringe was to face a cruel disappointment. He thought that, "as Tuscany had been the birth place of true Physics, the interest in this science would still be preserved, as in the times of Galileo, Torricelli and the Academy of Cimento" and that the lectures he had given in Lunéville would have even more success in Florence.[43] But, according to him, the young nobles of the town were involved in a totally different matter: they were dealing with a "kind of *galanterie*, named *sigisbéature*. It consists in spending part of the life around a woman, without being her husband, but serving her in every way".[44] Such intimate liaisons, together with the related expenses, did not allow the Florentine nobles, according to Vayringe, "to pay attention to the Program I published [...] My being a foreigner contributed a lot to

---

newtonien. Il y a pour environ dix mille écus de machines de toute espèce. Un simple serrurier devenu philosophe, et envoyé en Angleterre par le feu duc Léopold, a fait de sa main la plupart de ces machines, et les démontre avec beaucoup de netteté. Il n'y a en France rien de pareil à cet établissement"; Voltaire, Letter to de Formont, 25 june 1735, in Voltaire, *Oeuvres complètes*, Paris, 1835–1838, 13 vols.; v. 11, p. 159.

[43] "[Je m'étois figuré que] la Toscane ayant été comme le berceau de la véritable Physique, le goût à cette science s'y seroit conservé, comme au tems des Galilées, des Torricelles, & de l'Académie del Cimento, & que, par conséquent, les Leçons que j'avais données à Lunéville, auroient encore plus de vogue à Florence"; A. Calmet, op. cit. (n. 40), p. 997.

[44] "[...] une sorte de galanterie, que l'on qualifie du nom de *Sigisbéature*, Elle consiste à passer une partie de sa vie autour d'une femme, dont on n'est point le mari, & lui rendre tous les devoirs & les petits soins"; Ibid.

this indifference".[45] All this, he said, "has been the pitfall for the Academy of Lunéville, which was moved in Tuscany with immense expenses, and provided with the same Professors who had made it flourish; it [the Academy] is now desert; the Physics Room, one of the most original and complete in Europe, has suffered the same fate, though the price of the lessons has been reduced to less than a half than what was paid in Lunéville".[46] Very bitterly, he added that his skilfulness in mechanics, which in France would have led him to fortune, had thus become totally useless.

Through Agelli, Poleni at first asked Vayringe for three instruments, which he needed very urgently, as he wrote on 1 December 1741, "I do not see any other way (in order to get into the good graces of Monsieur Vayringe in time, as I need to) unless he gives me, as for those three little machines I asked, the ones he uses himself".[47] We do not know whether Vayringe accepted, but three instruments made by Vayringe are listed in Poleni's catalogue of entries at the very beginning of 1743. Satisfied with this first purchase, Poleni asked Vayringe for a device to study the refraction of light in liquids, together with a drawing of the instrument itself, which Poleni intended to use to enhance the quality of his explanations during his lectures.[48] Vayringe took quite a while – from July to October 1742 – to make the device and the drawing because, according to Agelli, "Our man [Vayringe] has thousands of occupations, and he had better carry out everything, as he has several unmarried daughters and no sons".[49] However, the instrument that was finally sent made Poleni forget his impatience, and the Venetian scholar personally wrote to Vayringe: "You write to me that one only needs to look at the machine for the refraction of fluids to understand how it works. And you have written the truth

[45] "[...] de faire attention au Programme que je publiai [...] Il est vrai que ma qualité d'étranger ne contribua pas peu à cette indifférence"; Ibid., p. 998.

[46] "[...] ont été l'écueil de l'Académie de Lorraine transférée en Toscane, avec des frais immenses, & pourvue des mêmes Professeurs qui l'avoient rendue florissante; elle y est devenue entiérement déserte; la Chambre de Physique, une des plus curieuses & des plus complettes qui soient en Europe, a eû le même sort, quoique le prix des Leçons que l'on y donnoit, ait été réduit à moins de moitié de ce que l'on payoit à Lunéville"; Ibid.

[47] "[...] non so vedere altro mezzo (per ricevere le grazie del Monsieur Vayringe in tempo, in cui potessero servirmi) se non che egli mi desse quelle tre machinette da me richieste, che egli tiene per suo uso"; Poleni, Letter to Agelli, 1 December 1741: BCVr, Carteggio Poleni, v. 3.

[48] Poleni had in his Cabinet several of such illustrated cardboards; Poleni, "Indice delle macchine", cit. (n. 32).

[49] "Il nostro signore ha mille occupazioni, e gli conviene por mano a tutto essendo carico di ragazze nubili e senza maschi"; Antonio Agelli, Letter to Poleni, 1 September 1742: BMVe, ms. it., IV, 643 (= 5504).

[. . .]. But this is precisely the great merit of your work [. . .]. Easiness is the most difficult part in this beautiful art: but it is the most useful as well".[50]

Poleni enthusiastically described his new instrument to Francesco Algarotti in 1746: "To measure the refractive powers of fluids, I have a very useful machine, and worked with a refined accuracy. So that I have the opinion (probably well-founded) that [. . .] no better machine can be found. With this, I studied the refractive power of cold water".[51] In 1743, Vayringe made three more instruments for Poleni but he fell ill in May 1744, and did not recover but died in 1746. This ended the fruitful collaboration between the two men, which had brought to Padua devices made by an instrument-maker coming from Lorraine but, here again, strongly inspired by the English and Dutch Newtonian style of lecturing (*Fig. 6*).

The Cabinet of Physics of Padua is thus a result of the circulation of instruments, scientists, makers and treatises in eighteenth-century Europe. As for Poleni, he actively contributed to this circulation: towards Portugal, through the work of Dalla Bella in Coimbra, and towards Greece, since some of Poleni's Greek students were among the scholars who introduced experimental physics in Greece.[52] Padua was thus totally involved in the spread of experimental philosophy lectures and physics collections that characterised the Enlightenment. It is however important to point out that Poleni's scientific interests and ideas, as well as the peculiarities of the Paduan and Italian context, provided the University Collection of Padua, as we have seen, with an identity of its own, so that Padua in turn enriched with peculiar and new elements the diffusion of experimental philosophy lecturing in Europe. More analyses still have to be carried out on Poleni's manuscripts and correspondence in order to examine further both the local situation and the circulation of scientific ideas and craftsmanship among European scientists and instrument-makers, as well as

---

[50] "Ella mi scrive non esservi bisogno che di riguardare la machina inserviente alla refrazione de' fluidi per intenderne l'uso. Ed ha scritto il vero [. . .]. Ma ciò è appunto quel che fa il gran merito alla di lei opera [. . .]. La facilità sì è la parte difficilissima in questa bell'arte: ma è l'utilissima ancora"; Poleni, Letter to Vayringe, 26 October 1742: BMVe, ms. it., IV, 643 (= 5504).

[51] "Io per misurare le potenze de' fluidi, rifrattive, ho una machina assai utile, e lavorata con una diligenza distinta. Onde ho anche un'opinione (forse ben fondata) che [. . .] [non] si possa ritrovare macchina migliore. Con questa ho cercata la forza rifrattiva di essa acqua fredda"; Poleni, Letter to Algarotti, 13 January 1746: BCVr, Carteggio Poleni, v. 1, ff. 127–128.

[52] George N. Vlahakis, "An outline of the Introduction of Classical Physics in Greece. The Role of the Italian Universities and Publications", *History of the Universities* (1998), pp. 157–180.

Fig. 6. Instrument for studying the refraction of light in liquids, made by Philippe Vayringe (Museum of the History of Physics, University of Padua).

the role and place of science within the intellectual, cultural, sociological and political context of the time. This should contribute not only to identifying the peculiarities of Padua and Italy, but also to enlightening the emergence of a broader European context as a whole.

THE LOST CABINET OF EXPERIMENTAL PHILOSOPHY OF THE UNIVERSITY OF OXFORD

Jim Bennett

The Cabinet

The building in Broad Street, Oxford, that is now the home of the Museum of the History of Science, is often said to be the original Ashmolean Museum, which was accommodated there from its foundation in 1683 till its move to Beaumont Street towards the end of the nineteenth century. This is only partly true. In fact the seventeenth-century project for the building was much more ambitious and inclusive than the Ashmolean today, embracing nothing less that the reform of natural knowledge in the manner associated in seventeenth-century England with the notion of "Solomon's House".[1] Francis Bacon described such an institution, which he called "Salomon's House", in his mythical tale of the island Bensalem – a "New Atlantis" – where shipwrecked travellers find a utopian society, whose success derived from a collaborative, institutionalised enterprise of experimentation, collection and documentation. This vision was influential in the founding of the Royal Society and of what, for want of a more official contemporary title, we might call the Old Ashmolean.

The Old Ashmolean had three floors with three complementary functions. The museum collections were at the top, a teaching room for experimental natural philosophy on the middle floor, and a chemical laboratory in the basement, which also accommodated at times anatomical dissections and a dispensary or pharmacy. Natural knowledge would be made in the basement and taught through lecture-demonstrations in the middle floor, while the natural world, and to a lesser extent the artificial, would be collected, classified, documented and displayed in a gallery at the top.

[1] Jim A. Bennett, Stephen A. Johnston, Anthony V. Simcock, *Solomon's House in Oxford: New Finds from the First Museum*, Oxford, 2000.

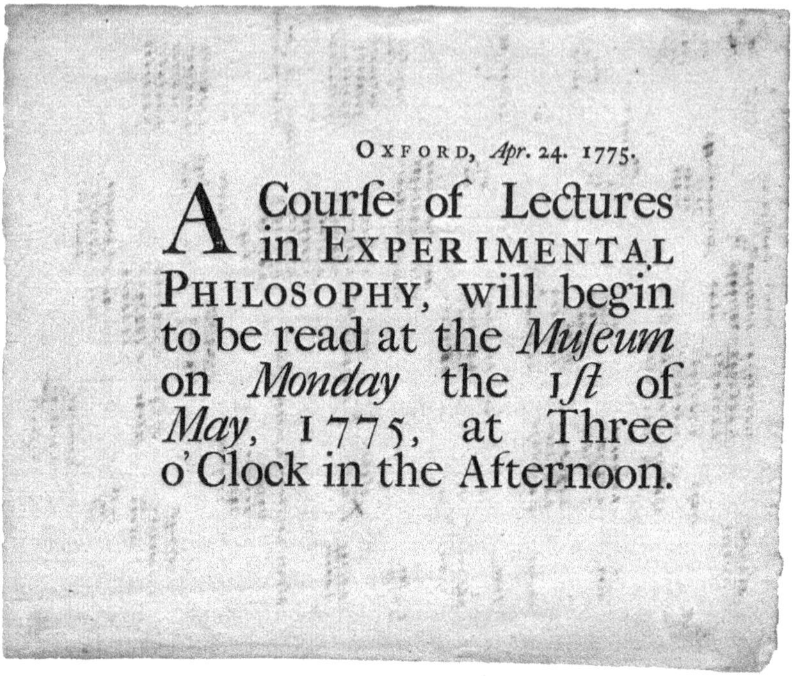

OXFORD, *Apr.* 24. 1775.

A Courſe of Lectures in EXPERIMENTAL PHILOSOPHY, will begin to be read at the *Muſeum* on *Monday* the 1*ſt* of *May*, 1775, at Three o'Clock in the Afternoon.

Fig. 1. One of a number of printed announcements of Hornsby's lectures, issued for different dates and now preserved at the Museum of the History of Science. The verso of this sheet has been used for calculations (Museum of the History of Science, University of Oxford).

Ashmole himself had anticipated the provision of teaching in the history of nature in the building and the middle floor became *Schola Naturalis Historiae*, or the "School of Natural History". John Keill and John Theophilus Desaguliers taught experimental natural philosophy elsewhere in Oxford in the early years of the eighteenth century, but this activity was brought into the School of Natural History by John Whiteside, the Keeper of the Ashmolean Museum from 1714. It was he who began the collection of instruments, many of them probably made in London by the younger Francis Hauksbee. This collection was purchased after Whiteside's death in 1729 by James Bradley, Keill's successor as Savilian Professor of Astronomy, and, although he did not become Keeper of the Museum, he continued the courses of lectures, extended the collection and was appointed Reader in Experimental Philosophy in 1749. Bradley was succeeded in both his posts by Thomas Hornsby in 1763, and Hornsby continued the now-established practice of regular lecture courses (*Fig. 1*), illustrated by experiments and demonstrations, while improving the collection at his own expense.[2] A traveller well able to judge, Thomas Bugge, visiting Hornsby in 1777, noted a "beautiful collection of physical instruments".[3]

It is from Hornsby's period, with the collection at its finest and most extensive, that we have the best record of its contents. Several inventories from Hornsby's time give us a very detailed account of this lost collection. The one example that has been noted in the existing literature on the history of the building dates from 1790.[4] This inventory has been attributed to the instrument maker Edward Nairne but, although he was involved with the project for an inventory, he does not seem to have been its author. It is very full, listing not only complete, self-contained instruments, but also the kind of loose apparatus that might well be found in the context of lecture-demonstrations. So, as well as entries such as "The Armillary Sphere to shew the diurnal Motion 18 Inches diameter", there are others of the character, "Large tubes terminating in a Capillary bore, capillary tubes of different bores in a piece of Cork".[5]

The inventory has about 330 entries, many of which refer to a number of items – "Drawer with artificial Magnets about 8 or 10", "Four tin Machines with Threads to represent the passage of Rays thro Lenses" or a pair of globes.[6] About 160 entries relate to coherent instruments for a particular purpose, rather than miscellaneous pieces of apparatus (in the nature of things, the figure is approximate). It seems likely that the miscellaneous pieces did had a specific purpose, such is the structured nature of the collection, but we do not always know what this was, the functions of the odd pieces being mentioned in only a few cases, such as "Two Tubes for the Torricellian Experiment".[7] While others seem less purposeful, such as, "Pieces of Copper, Lead & brass with strings",[8] the context in which they appear can help to make some sense of them, for the whole correction

---

2 For the lecturing activities in the School of Natural History, see Ibid., pp. 18–20. Inga Elmqvist Söderlund has discovered a previously-unnoticed first-hand account of a lecture by Bradley on mechanics, involving the use of instruments, witnessed by a Swedish visitor to Oxford, Bengt Ferrner, on 7 March 1760; private communication.

3 Kurt Moller Pedersen (ed.), *Thomas Bugge, Journal of a Voyage through Holland and England, 1777*, Aarhus, 1997, pp. 212–217.

4 Anthony V. Simcock, *The Ashmolean Museum and Oxford Science*, Oxford, 1984, p. 13.

5 *A Catalogue of the Philosophical Apparatus belonging To The Rev^d: D^r: Hornsby of Oxford*: Oxford, Bodleian Library (Bodl.), MS Top. Oxon.c.236, ff. 3–13, see ff. 7, 5.

6 Ibid., ff. 5, 7v, 7.

7 Ibid., f. 9.

8 Ibid., f. 8v.

is divided into 26 "Days", which refer to the sequence of lectures and surely also to the arrangement of cupboards in the room.

As curators and instrument enthusiasts, we will regret the loss of the collection – not a single piece can be identified with any confidence – but as historians we can be grateful for a very full record of this significant instance in the narrative of the cabinet of experimental philosophy. What we lack in objects is generously compensated for by very rich documentation. The value of the Oxford cabinet is that we can place the instruments in a very well-documented account of the teaching they supported, see where they came into play in lectures and how they were deployed.

The 1790 inventory records many of the classic instruments of experimental philosophy. There are an electrical machine, air pumps, "Guerricks Hemispheres", orreries, "Two Models of the Lifting Pump", "Two Models of the squirting Pump", "the Pile Engine", "The Crane", "The Corn Mill with a Water Wheel", "The Coal Gin (very large)", "Machine to Illustrate the different strength of Horses", "The large Machine to Illustrate the Wedge", and so on. The inventory continues at length in a fairly familiar way but there are also some unusual or special items, such as "Wooden Table with 32 Compasses to Illustrate Dr Halley's Hypothesis" or "Ferguson's Machine to shew that the path of [the] Moon is concave to the Sun". Only very rarely are makers mentioned: a 4-foot achromatic telescope by Dollond and a pair of 18-inch globes by Adams. We might be more interested to know the maker of two "old 12 Inch Globes, Celestial & Terrestrial" but, since they are old, they are not so interesting to the compiler of the inventory.[9]

The character of the collection is clearly didactic – its purpose is teaching – and this applies even to the instruments that would have quite a different role in a different setting – practical ones like an azimuth compass, a variation compass, or a "Sea Compass on Gimbals", or regular items of instrument commerce like seven barometers, covering a range of types. Here they are for learning natural philosophy through illustration and demonstration. We can be sure of that because we have the text for that demonstrative teaching, which gives us a close, parallel narrative.

Before moving to this narrative, however, we should ask what the 1790 inventory was for: administrative housekeeping, for example, checking the collection from time to time and keeping it in order? The fact that it had a more specific function is revealed at the end of the list, where there

---

[9] Ibid., *passim*.

is a valuation of the whole collection by Nairne. His initial figure is £368 7s 6d, but then he has to add 7 guineas, because he has forgotten to include the air-gun in the total, which is revised to £375 14s 6d.[10] We do not have Nairne's working in reaching this total, so we lack his valuations of the individual pieces, except in the case of the air-gun.

Why was a valuation needed? Previously the collection had belonged to the professor. We know that Bradley bought Whiteside's instruments and added to them; Hornsby, who had himself attended Bradley's lectures,[11] certainly made further additions and the content of the inventory implies that he had also bought Bradley's instruments in turn. What had changed now was that the University had been given a benefaction to support and consolidate the teaching of experimental philosophy and it was beginning to spend it.

Following the inventory, a different hand turns the manuscript into a legal document, where to Nairne's valuation of Hornsby's apparatus is added the charge for the valuation (10 guineas) and the valuation of the cupboards and cases (valued not by Nairne, but a Mr Pears).[12] So we know the instruments were in cupboards, no doubt around the walls of the School of Natural History. The total sum is given to Hornsby, implying that he owned the cupboards and that he had engaged Nairne. The legal agreement seems to make a distinction between the instruments of Bradley and Bliss, who had been Professor of Geometry, and the apparatus of Hornsby, so there remains a lingering doubt over whether Hornsby did purchase Bradley's instruments and so was being paid for the whole cabinet at this stage. Hornsby signs the document, agreeing, for example, that he will not buy any more instruments without the prior consent of the trustees of the fund, and last of all is Nairne's signed acknowledgement of his fee, which he did not receive until June the following year, 7 months later, presumably from Hornsby.

---

10 Ibid., f. 11.

11 Thomas Hornsby, "Memorandums of Dr. Bradleys Course of Experiments": Bodl., MS Rigaud 70. Hornsby occasionally refers to Bradley's work in his lectures: Hornsby, "Laws of Motion": Bodl., MS Radcliffe Trust d.9, f. 216; Hornsby, "Introduction to Astronomy": Bodl., MS Radcliffe Trust d.16, f. 132a; Hornsby, "Light": Bodl., MS Radcliffe Trust d.11.

12 *A Catalogue*..., cit. (n. 5), ff. 11–12.

The Lectures

Moving beyond the 1790 inventory, what more can we discover about the use of the collection in the Old Ashmolean? An outstandingly rich resource is Hornsby's very full lecture notes: written-out lectures in multiple, redrafted versions. There are 16 substantial manuscript volumes, generally containing lectures written in Hornsby's hand, word for word as they were to be delivered, interrupted from time to time by a reference to an experiment or a short sequence of experiments.[13] We see how the use of instruments occurred within the lectures and exactly the points they were intended to illustrate.

We shall extract some instructive examples from four of these volumes for this limited study, but the pattern they reveal of talk interrupted by experiments will surely be consistent throughout. Hornsby did not publish his lectures, which makes the manuscript record all the more important. The beginning of the whole course is perhaps worth recording here, as an authentic contemporary introduction to the discipline of experimental natural philosophy. Had an auditor enrolled for the lectures in around 1770 and come to the School of Natural History to listen to Professor Hornsby, they would have heard these encouraging words:

> The business of Natural Philosophy is to describe the Phænomena of Nature; to give some Account of their Causes so far as they depend upon the known Properties of Bodies; & to ascertain those general Laws of Nature, according to which the Motions & Actions of Bodies are regulated. A strong Curiosity has prompted Men in all times to study Nature; Every useful Art has some Connection with this Science; & the Beauty & Variety of Things makes it ever agreeable & surprising.[14]

"Ever agreeable & surprising" sets the pleasant and engaging tone that Hornsby maintains throughout his course.

Hornsby sometimes reflects on the history of the subjects he takes up and gives us insights into how the past of science was viewed and taught in the eighteenth century. In the broad sweep of his introduction, he is very confident about what has been achieved in the previous two centuries and very dismissive of almost everything before then.

---

[13] The volumes referred to are in the papers of the Radcliffe Trust (see references in note 11 and below) but note also Bodl., MS Rigaud 54, ff. 188–329.

[14] Hornsby, "Laws of Motion", cit. (n. 11), f. 30 and f. 27.

> [The] Progress [...] which the knowledge of Nature made in many former Ages is exceedingly inconsiderable when compared with the numberless improvements it has received from the Discoveries made in the Course of the last Two Centuries only: insomuch that some of the Branches of Natural Philosophy, which at present is almost complete in all its parts were utterly unknown 200 Years ago. The reason of this seems upon proper Enquiry to be justly attributed to the practice of the Philosophers of former times; who in their pursuit after Natural Knowledge disregarded Experiments & busied themselves only in framing Hypotheses to explain in what manner any particular Appearance might have been produced.[15]

He begins with mechanics and quickly establishes a pattern that is carried through the lectures, one of exposition interrupted by notes such as "Proved by Exp^{t}",[16] or "explained by the Diagonal Instrument".[17] A few pages further on there is a more extended experimental interlude: "The Truth of what has now been advanced may be illustrated by several Experiments", whereupon he lists four:

> 1. By the Motion of a Brass Square in an horizontal direction...
> 2. By the Brass Diagonal Machine
> 3. Let three Ivory Balls of equal size be suspended from three Pins by strings of equal lengths...
> 4. By the Diagonal Jet of the Fountain.[18]

Later still, he announces, "I shall now proceed to illustrate & prove what has already been said by such experiments as may be proper for this purpose",[19] the list including the "Whirling Table", of which he seems to be fond, and adds a note to himself, "Shew M^{r}. Cantons Experiment of the two Whirligigs supported by an artificial Magnet".[20]

There is no need to multiply examples but the lectures are regularly punctuated with these experimental asides. One worth recording, perhaps, is Hornsby's introduction to his first lecture on the orrery: "Let us now take a general View of the Universe."[21] The orrery lectures reveal another of Hornsby's teaching techniques: he seems to have charts or large diagrams as well as instruments. He explains that the orrery can deal with relative motions but not with distances or magnitudes:

15 Ibid., f. 32.
16 Ibid., f. 64v.
17 Ibid., f. 73v.
18 Ibid., f. 93, *cf.* f. 121.
19 Ibid., f. 105, *cf.* f. 128.
20 Ibid., f. 105.
21 Ibid., f. 198.

> The Proportional Distances are in the Scheme before you, their Magnitudes are also before you; but neither the proportional Distances, or Magnitudes can be conveniently express'd in an Instrument of this Sort.[22]

When dealing with the detection of parallax by double star measurement, he says simply, "see the Scheme."[23] In general, then, thanks to the lectures, we can locate where the instruments in the 1790 inventory were deployed in Hornsby's natural philosophy.

The lectures offer other insights into eighteenth-century attitudes to instrumentation. Hornsby says, for example, that the electrical machine is now generally made with a glass cylinder instead of a globe, though some have been made recently with one or two circular plates.[24] One machine with two plates was particularly worthy of note:

> One of this Form was made in England for the Museum of Teyler at Harlem. It consisted on two Circular Plates of 65 Inches in diameter fixed on a common Axis parallel to each other and 7½ Inches asunder.[25]

This corresponds to the famous large generator in Haarlem – made by an Englishman, John Cuthbertson, but not in England.

Hornsby tells the story of Musschenbroek in Leiden in 1746, "much attended to & frequently repeated by all Electricians, when an Accident gave birth to a very remarkable [experiment], which has since engaged the Attention of every Lover and Promoter of this branch of Knowledge".[26] It resulted in "a disagreeable sensation; for which reason the Effect has been called the Electric Shock".[27] Hornsby relates the electrical experiments of the Abbé Nollet in much the same terms as we still use at present, and the discovery by Henry Gellibrand of the variation of the magnetic variation, very much as students are still told when learning the history of science in the same building today.[28]

Just occasionally Hornsby's accounts of the achievements of others come right up to date, as with the discovery of what he calls the Georgian Planet, and we call Uranus:

---

22 Ibid., f. 202.
23 Hornsby, "Introduction to Astronomy", cit. (n. 11), f. 178.
24 Hornsby, "Electricity": Bodl., MS Radcliffe Trust d.15, f. 7v.
25 Ibid., f. 7v.
26 Ibid., f. 10v.
27 Ibid., f. 18.
28 Ibid., *passim*.

> From these considerations therefore it should seem that M^{r}. Herschel has had the happiness to discover a new Planet revolving round the Sun [...] & be allowed to increase the Number of those Bodies, which from the earliest period of time to the present Year have received no addition, & equalled only in Number the 7 days of the Creation.[29]

Hornsby kept lists of those attending the various courses and quite a few are preserved in the volumes of lectures.

One interesting fact we learn from Hornsby's text is that Bradley made an unexpected use of part of the famous zenith sector he commissioned from George Graham and with which he detected the aberration of light, the first observational proof of the orbital motion of the Earth. Hornsby tells his students that for Bradley's investigations of water flowing from a cistern, he measured the diameters of carefully-prepared brass jets "by help of the Screw belonging to his Parallactic Instrument".[30] By comparison with measurements with a micrometer, Bradley had estimated that he knew the diameters of each of his jets to one thousandth of an inch.

One of the lecture volumes contains another inventory, this one in Hornsby's hand, and it is practically the same as the 1790 listing.[31] It is so close that one is virtually a copy of the other: the word-for-word similarities in the long descriptions show that at the very least one list was to hand while the other was being compiled. It seems that the 1790 list is the copy, since it has a few mistakes: most notably the air-gun that had to be added to the valuation at the last minute is not in the 1790 inventory but is in the Hornsby list. So, the 1790 inventory was not compiled by Nairne at all, as has been assumed. In fact, if it is read carefully, only the valuation is attributed to him and not the inventory. As he went through the collection in the School of Natural History, Nairne was working from a copy of Hornsby's inventory in the hand of an amanuensis, which was missing the air-gun entry, an error probably pointed out by Hornsby, as he was the one being paid for the collection.

There is, in fact, a third list (at least a third; there may be others), which is more of a working notebook than a formal inventory.[32] This is something Hornsby had been using, probably relatively early in his lecturing career, to keep track of items necessary for illustrating his lectures and

29 Hornsby, "Introduction to Astronomy", cit. (n. 11), f. 209v.

30 Hornsby, "Laws of Motion", cit. (n. 11), f. 32; Hornsby deals with Bradley's attempt to measure stellar parallax at Hornsby, "Introduction to Astronomy", cit. (n. 11), f. 132a.

31 Hornsby, "Electricity", cit. (n. 24), f. 189 ff.

32 Hornsby, "[List of items]": Bodl., MS Radcliffe Trust e.11.

recording some of their movements. There is a pattern where for each of the 26 days, there is a first page of entries marked “Mine”, then a page with the number of the day repeated but without “Mine”, and then a few blank pages. Items that are “mine” were perhaps added to the Bradley instruments by Hornsby himself. The whirling table was “mine” as were the Adams globes.[33] Several “Animals” are “mine”, and they would surely have been acquired later than Bradley’s pieces and were intended for the lecture on respiration.

We can conclude with an example of how the lectures can illuminate the entries in the inventories, and so the use of types of instrument in lectures generally. The 1790 inventory includes an inconspicuous item, “Metal Cylinder with its Pictures”,[34] among optical bits and pieces. It is obviously a cylindrical mirror with anamorphic drawings or prints, and it might be considered out of character with a teaching collection, being little more than an optical toy.

If we then turn to Hornsby’s lectures on optics, we find that the pattern of his practice – the use of diagrams and experiments – is the familiar one, with notes to himself like: “Shew the Scheme & prove by Exp^{t}”.[35] He first seems to agree with the idea that the cylindrical mirror is only a toy:

> Cylindrical, Conical, Prismatical & Pyramidal Mirrors are to be considered as mere Curiosities. They give irregular Images of Regular Objects presented to them; or they give a Regular Image of an Object, provided the degree of Irregularity be accommodated to the Circumstances of the Mirror.[36]

However he goes on to give an analysis of the distortion produced by cylindrical and conical mirrors, and then to an exercise in, as he says:

> the Manner of designing upon an horizontal Plane a disfigured Object, so that by placing Vertically a Cylindrical Speculum upon that Plane of a given radius, that object may appear Vertical and Regular to an Eye in a given situation.[37]

The treatment is quite mathematical, at least for Hornsby’s lectures, which do not really have much mathematical content. So here is a case where an instrument can be given an alternative context and significance by the extensive documentation for the lost collection that extends our understanding of how its contents were used.

---

33 Ibid., ff. 4v, 5.

34 *A Catalogue*..., cit. (n. 5), f. 7v.

35 Hornsby, “Light”, cit. (n. 11), f. 24.

36 Ibid., f. 27.

37 Ibid., f. 28.

FAILURE, FRAUD AND INSTRUMENT CABINETS: ACADEMIC INVOLVEMENT IN THE EIGHTEENTH-CENTURY DUTCH WATER CRISIS

Tiemen Cocquyt

In 1788, the newly appointed professor of physics and mathematics, Christiaan H. Damen, was granted permission from the Leiden University Board to dispose of those instruments that were of use "neither for the study of the history of machines, nor for the instruction of physics".[1] If it was Damen's goal to maintain a well-equipped academic cabinet for physics instruction, he was at the right place. A setting for vacuum and mechanical experiments since the 1670s, Leiden University had been among the earliest academic institutions to introduce instruments into its philosophy curriculum.[2] This tradition was brought to further heights when, in the first half of the eighteenth century, the professors Willem J. 's Gravesande and Pieter van Musschenbroek systematized empirical demonstration in their natural philosophy textbooks, and thereby provided a means for the diffusion of Newtonian philosophy on the Continent.[3] The canon of demonstration apparatus they had brought together, enlarged with acquisitions by their successors, was an enviable collection for anyone wishing to enrich their lectures with experimental demonstration. Why was it, then, that Damen felt the urge to clean out parts of this cabinet which had been formed, as he stated himself, "in the cradle and nursery of experimental physics"?[4] As Peter de Clercq has shown, Damen's move can be understood by taking into account the declining position the Leiden Cabinet of Physics was in during the decades following its "highlight period"

[1] Peter de Clercq, "The 's Gravesande Collection in the Museum Boerhaave, Leiden", *Nuncius: Annali di Storia della Scienza* 3 (1988), pp. 127–137: 133.

[2] Edward G. Ruestow, *Physics at Seventeenth and Eighteenth-Century Leiden*, The Hague, 1973, pp. 96–112; Gerhard Wiesenfeldt, *Leerer Raum in Minervas Haus: experimentelle Naturlehre an der Universität Leiden*, Amsterdam, 2002.

[3] Ruestow, op. cit. (n. 2), pp. 113–139. Also see the article of Hans Hooijmaijers and Ad Maas, *infra*, pp. 27 ff.

[4] P. de Clercq, "In de schaduw van 's Gravesande. Het Leids Physisch Kabinet in de tweede helft van de 18de eeuw", *Tijdschrift voor de Geschiedenis der Geneeskunde, Natuurwetenschappen, Wiskunde en Techniek* 10 (1987), pp. 149–173: 149.

under 's Gravesande.[5] In trying to keep up with new developments, there were strategic reasons for Damen to stress the need for innovation to the University Board, while at the same time emphasizing the significance of his predecessors' achievements. What Damen was doing was shaping the history of his predecessors' research, and – at a material level – the cabinet and the objects they had brought together.

In this article, I want to sketch some events that preceded Damen's actions, and investigate how, already during the formation of the historic cabinet, Leiden professors disseminated their activities along a well-considered strategy. I will set out how the academic authority of the Leiden professors, in an era of growing public awareness for experimental philosophy, was not an immutable achievement, but instead was something that needed to be maintained actively. In keeping up – particularly in their hydraulic investigations – the professors involved themselves in the complex interplay of state affairs, politics, invention and entrepreneurship. Yet the risks and vulnerabilities, intrinsic to such interplay, are not always included in the textbooks and publications the professors authored. The shaping of stable, systematized experimental philosophy from risky enterprises was a process that was already going on in the times before Damen made a clean sweep of the Physics Cabinet. In the second part of this article I want to illustrate what repercussions this has for instrument cabinets. It will be argued how a more dynamic view of the interaction between public and academy, rather than a one-way model of science dissemination, gives us a different understanding of some of the Physics Cabinet's "demonstration" instruments.

's Gravesande and the Wheel of Wonder

In 1721 's Gravesande spent the summer at the court of the Landgrave Carl of Hessen-Kassel. Contacts between rulers of the German lands and the Leiden experimental community had existed for a longer time; at least since the 1680s, visits of rulers and professors to the Musschenbroek workshop had enriched the German knowledge centres with philosophical instruments and references to them.[6] The court of Hessen-Kassel had been particularly active in such exchange of knowledge and technology.

[5] Ibid., pp. 171–172.

[6] P. de Clercq, *At the Sign of the Oriental Lamp: The Musschenbroek Workshop in Leiden. 1660–1750*, Rotterdam, 1997, pp. 152–163.

At the same time, the Landgrave was initiating construction works for the *Bergpark Wilhelmshöhe*, a monumental hillside park which was to surpass the splendour of even the largest courtly gardens of the time. The highest point of the park was reserved for the impressive *Hercules* statue, with water cascading down from it along the gigantic water-stairs. As an illustration of early instruments not being deployed for mere philosophical speculation, it is interesting to note that at Wilhelmshöhe, initial attempts to raise the water to higher levels were performed with a Senguerd-type air-pump, manufactured in 1686 by the Leiden Musschenbroek workshop.[7] Unfortunately, raising the water above the natural limit of circa 10 metres proved impossible, and other solutions needed to be reverted to. Nevertheless, opportunities to achieve fame and profit with hydrographical inventions remained present in Kassel in the following decades. It is against this background that we must situate the contrivance of Johann Ernst Elias Bessler, also known as "Orffyreus".[8] Bessler's device was a perpetually moving wheel that furthermore could lift a mass, a bucket full of water for example, without any loss of movement (*Fig. 1*). Bessler jealously guarded the working principle of the machine, but was always willing to sell the secret for a vast amount of money. Earlier versions of the wheel had been built at other German courts but now, in 1717, the Landgrave offered Bessler a forum to construct his largest wheel so far. It did not take long for rumours about the invention, both positive and critical, to spread all over the Continent. In order to elucidate the matter and lay the case to rest, the Landgrave invited 's Gravesande to give his opinion on the machine.

The choice for 's Gravesande, a mediator of experimental philosophy and early technology, was not at all surprising. In 1715, joining a diplomatic mission to England, he had been introduced to the Royal Society and its president Isaac Newton.[9] Also during this stay, together with his British

[7] Karsten Gaulke, "Experimentelle Naturlehre und Gartenkunst: Was haben Vakuumpumpen und Dampfmachinen mit den Wasserkünsten zu tun?", in Sandra Kress, Jennifer Verhoeven (eds.), *Hortus ex Machina. Der Bergpark Wilhelmshöhe im Dreiklang von Kunst, Natur und Technik*, Wiesbaden, 2010, pp. 156–166.

[8] 's Gravesande's involvement with Bessler's invention is described extensively by his biographer: Jean N.S. Allamand, "Histoire de la vie et des ouvrages de Mr. 's Gravesande", in Allamand (ed.), *Oeuvres philosophiques et mathématiques de Mr. G.J. 's Gravesande*, Amsterdam, 1774, pp. ix–lix. See also: Claude A. Crommelin, "La roue d'Orffyreus", *Janus* 48 (1960), pp. 47–52; Simon Schaffer, "The Show that Never Ends. Perpetual Motion in the Early Eighteenth Century", *British Journal for the History of Science* 28 (1995), pp. 157–189.

[9] The biographical information that follows has been taken from Allamand, "Histoire de la vie…", cit. (n. 8), pp. xxii ff.

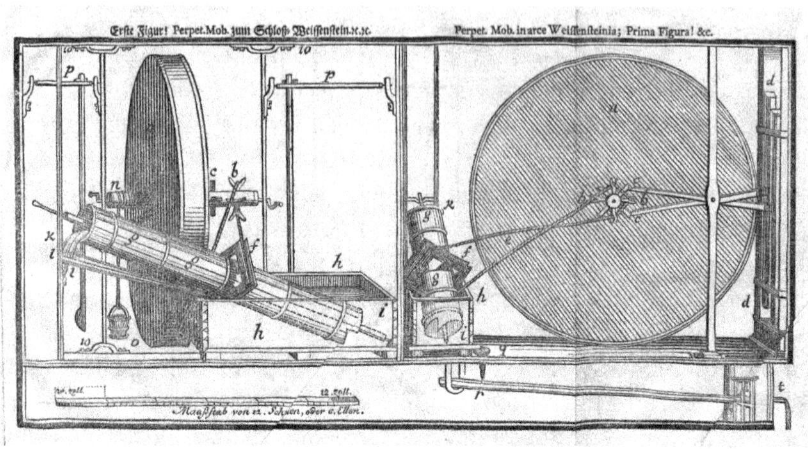

Fig. 1. Engraving of the perpetual motion Bessler built at the Landgrave's hunting lodge; Johann Ernst Elias Bessler, *Das triumphirende perpetuum mobile Orffyreanum*, Kassel, 1719 (Utrecht University Library, sign. P qu 752).

congener John Theophilus Desaguliers, he had worked and improved on the Savery steam engine. Now, in 1721, during a summer stay for discussing the introduction of steam technology in Hessen, 's Gravesande inspected Bessler's wheel and was truly astonished. As he would later recall, he found himself in an "amazement impossible to describe".[10] There was nothing that hinted at any form of fraud, and he held the mechanical and mathematical talents of Bessler in such high regard that the machine had to be real anyway. Even better, the machine offered such unprecedented possibilities that it would be a shame to keep these unexploited for the public, and for the advancement of science as a whole. Once returned to Leiden, he promptly sent a letter to Newton asking for an appropriate reward for Bessler.

Much to his dismay, things did not proceed as 's Gravesande had hoped for. Nobody in England felt compelled to provide the financial resources for Bessler's reward. Instead 's Gravesande's letter to Newton appeared in print, giving his enthusiasm for the perpetually moving wheel widespread attention.[11] At the same time Desaguliers demonstrated to the Royal Society that all devices in which overbalancing wheels remain in motion – Bessler's

---

[10] Willem J. 's Gravesande, "Lettre à Mr. Neuwton", *Mercure historique et politique* (September 1721), p. 363; reproduced in Allamand (ed.), *Oeuvres*…, cit. (n. 8), pp. 303–304: 303.

[11] Ibid., pp. 303–304.

wheel in particular – were founded on the wrong principles, and would never work.[12] The Leiden professor had made himself look foolish. But the events did set 's Gravesande thinking about perpetual motion and the forces of bodies in motion. What were these latter anyway? Translating the properties of rising and falling weights, the principles of which dominated the perpetual motion debate of the time, into an experiment with brass spheres and a tray of clay, 's Gravesande derived that the identification of force with the Newtonian "quantity of motion" had been wrong after all. A discussion of the background of 's Gravesande's find and the heated reactions it caused would be beyond the scope of this article.[13] What 's Gravesande did, in short, was to refute Newton on precisely those grounds that the English scientist propagated so strongly: empiricism. On the other hand, what the experiment also demonstrated was that the explanation for Bessler's wheel that the Leiden professor had derived, might not be valid after all. But despite this, 's Gravesande went to great lengths to stress that he did not rule out the possibility of perpetual motion completely.[14] Even in more recent historiography, 's Gravesande's adherence to the wonderful principle has been met with surprise.[15] For us, it is important to realise that the clay experiments 's Gravesande described in his 1722 *Essai d'une nouvelle theorie du choc des corps* were a direct consequence of the perpetual-motion considerations he published in his *Remarques touchant le mouvement perpétuel* shortly before.[16] It was during the construction of canonical experimental philosophy that the original order of events was lost. In the 1725 edition of the Leiden professor's *Physices elementa mathematica,* which incorporated the newly-devised experiment on the forces of motion, the references to the perpetual motion discussion had disappeared.[17] And while 's Gravesande's biographer Jean N.S. Allamand – about whom more is to follow – described the Bessler affair at length in his 1774 biography, we may suspect him of comparable dissociative

---

12 John Th. Desaguliers, "Remarks on some attempts made towards a perpetual motion", *Philosophical Transactions* 31 (1720–21), pp. 234–239.

13 See Thomas L. Hankins, "Eighteenth-Century Attempts to Resolve the *Vis viva* Controversy", *Isis* 56 (1965), pp. 281–297; Schaffer, op. cit. (n. 8), pp. 170–181; Jonathan I. Israel, *Enlightenment Contested: Philosophy, Modernity, and the Emancipation of Man 1650–1752*, New York and Oxford, 2006, pp. 201–222.

14 's Gravesande, *Remarques touchant le mouvement perpétuel*, 1722; reproduced in: Allamand (ed.), *Oeuvres*..., cit. (n. 8), pp. 305–312.

15 See, for example, Crommelin, op. cit. (n. 8), pp. 49–52.

16 's Gravesande, *Essai d'une nouvelle theorie du choc des corps*, The Hague, 1722; 's Gravesande, *Remarques*..., cit. (n. 14).

17 's Gravesande, *Physices elementa mathematica*, Leiden, 1725.

intentions in his insertion of the 1722 perpetual-motion considerations at the complete end of 's Gravesande's works, isolated from the contemporary publications on force experiments.[18]

Hydraulic Enterprises

To return to the Leiden experiments, in 1722 's Gravesande had found natural philosophical truth in a tray filled with clay. In the Netherlands of the early eighteenth century, clay and experimental philosophy were starting to intertwine more often. Despite the iconic role of the windmill for early Dutch energy production, the real motor behind the many farms, breweries, and workshops had traditionally been peat. The peat harvest, though not deployable directly for mechanical work, yielded over a hundred times as much energy as windmills in keeping the early-modern Dutch economy running.[19] This was mainly due to the Republic's strategic position: bogs could easily be harvested close to the urban centres, or be economically transported along the Dutch river deltas. For the latter, the sixteenth century saw the formation of powerful peat companies, which organised the harvest and the transportation in peat boats. When economically appropriate, new channels were dug to improve the supply system. The large scale grubbing was, however, not without consequence. By the late-seventeenth century the Dutch landscape had witnessed a considerable transformation. Natural river beds had diverted, and all the harvesting made existing water pools grow tremendously in size. The threat was so omnipresent that the waters in between the economic centres of Amsterdam, Leiden and Haarlem, the Haarlemmermeer, were coined the "water wolf", reflecting the dangers that it posed when it threatened the city gates during stormy weather.[20]

Traditionally, water management in the Netherlands had been the duty of the regional district water boards. But now, the scale and complexity of the problems demanded a cross-border approach. In their attempts to solve the water crisis, from the 1720s onwards, the States of Holland increasingly involved academics, one of whom was 's Gravesande. But

[18] Allamand (ed.), *Oeuvres*..., cit. (n. 8), pp. 305–312.

[19] Jan W. de Zeeuw, "Peat and the Dutch Golden Age: The Historical Meaning of Energy-Attainability", *AAG-bijdragen* 21 (1978), pp. 3–31.

[20] Cesare S. Maffioli, "Italian Hydraulics and Experimental Physics in Eighteenth-Century Holland", in Cesare S. Maffioli, Lodewijk C. Palm (eds.), *Italian Scientists in the Low Countries in the XVIIth and XVIIIth Centuries*, Amsterdam, 1989, pp. 243–275: 252–261.

he was not the only one; the advising milieu he was in was grafted onto Leiden University in other ways as well. Many other advisors, like Nicolaas Cruquius and Melchior Bolstra, had Leiden University as their *alma mater*. The advisor network 's Gravesande became involved with was a colourful group of people. It was a circuit in which experts in both practical and scholarly matters, politicians, and investors complemented each other in their attempts to resolve the water crisis. In recent times, the importance of engineering projects for the construction of Newtonian science as a public good in Britain has received proper attention.[21] As the events sketched in the following paragraphs will show, the situation in the Netherlands was not significantly different. What they testify to, in addition, is how the professors could gain, but also set at stake, their academic authority.

Yet sometimes, we must resist the urge to see only pure application as the underlying motive in scholars' engagement with steam-driven or water-elevating devices. As we have seen, 's Gravesande had worked on the Savery steam pump during his visit to England in 1715–1716, and it did not take very long for him to introduce a copy of this machine into the University cabinet. Records show that 's Gravesande ordered a Savery machine at the Musschenbroek workshop in 1727.[22] While not much is known about this early instrument, a similar acquisition by the Utrecht professor Pieter van Musschenbroek (*Fig. 2*), brother of the instrument maker Jan, illustrates that steam pumps gained a wider acceptance in Dutch academic circles in these years. Pieter ordered his pump around 1730 from a certain J.S. Meijer in Amsterdam.[23] It figures prominently in the 1739 edition of his *Beginselen der Natuurkunde*. What is striking is that by this time the Savery pump had already been superseded for a while by the more reliable and efficient, yet hybrid, Newcomen engine. The Dutch professors, it appears, appreciated the Savery pump primarily for the singular natural philosophical principle that lay at its foundation. Full of enthusiasm Pieter van Musschenbroek wrote about his device that "a small amount of vapour makes a jet of water jump 40 feet, even 50 feet high".[24] With respect to gunpowder, water vapour was "incomparably

[21] Larry Stewart, *The Rise of Public Science: Rhetoric, Technology and Natural Philosophy in Newtonian Britain, 1660–1750*, Cambridge, 1992.

[22] De Clercq, "In de schaduw…", cit. (n. 4), p. 163.

[23] Ibid., pp. 163–165.

[24] Pieter van Musschenbroek, *Beginsels der natuurkunde*, Leiden, 1739, p. 438.

Fig. 2. Savery steam pump by J.S. Meijer, c. 1730 (Museum Boerhaave, inv. no. V09650).

more powerful".[25] In contrast with these marvellous phenomena, real world problems, it seems, were to be solved not with steam but with proven, mechanical technology.

In this respect, it should also be noted that the English innovations in steam technology could not be applied right away to the water problems the Republic faced. When 's Gravesande listed the advantages of steam technology he could refer to "deep mines", but in the Netherlands these hardly existed.[26] What was ubiquitous were situations where large quantities of water needed to be elevated to heights with a maximum of 16 feet. Existing solutions with series of four polder mills were rather cumbersome.[27] Now, administrators, scholars and inventors were rowing in the same direction. Spurred by the water crisis, they opened the arena for hydraulic innovation.

An early attempt to come up with a solution for the limited elevation heights was proposed by the Danzig-born scientist Daniel Gabriel Fahrenheit. Residing in Amsterdam from 1717 onwards, this lecturer and developer of the mercury thermometer soon became friends with 's Gravesande. In the 1730s Fahrenheit developed a device that would solve the limited elevation height of traditional scoop wheels (mill-driven wheels used to lift water from one level to another).[28] His *waterbuysmolen* (literally, "water tube mill") consisted of an axis around which tilted tubes were fitted. The centrifugal forces created by rotating the machine would lift the water. Only a brief description of the apparatus survives, though it is interesting to note that 's Gravesande was improving on his cabinet centrifugal forces instrument in the same years.[29] Fahrenheit was given a patent for his machine by the States of Holland in 1736, but did not live to see his invention being put to work.[30] Shortly before his death he transferred the patent rights, and an experimental prototype of the machine, to 's Gravesande.[31] The Leiden professor subsequently put the principle to test and tried to "streamline" the design, but eventually abandoned it.

---

25 Ibid., p. 440.

26 De Clercq, "In de schaduw...", cit. (n. 4), p. 163.

27 The challenges are well enunciated in Johannes Lulofs, *Korte beschryving van het nieuwe water-rad, uitgevonden door Leopold de Genneté*, Dordrecht, 1756.

28 Gerard Doorman, *Octrooien voor uitvindingen in de Nederlanden uit de 16e–18e eeuw*, The Hague, 1940, pp. 308–309.

29 De Clercq, "The 's Gravesande Collection...", cit. (n. 1), pp. 130–131.

30 Doorman, op. cit. (n. 28), p. 309.

31 Lulofs, op. cit. (n. 27), p. 48.

The principle was sound, but in practice friction forces could not be overcome sufficiently.

For advocates of experimental philosophy like 's Gravesande and Van Musschenbroek, the 1730s must have been exciting times. It has been observed that this was the period in which Dutch public interest for experimental philosophy was blossoming.[32] This rise in interest also resulted in a growing number of inventors settling in the Republic, who tried to solve the water crisis in highly original manners. It would be an overstatement to say that all of these attempts were without success. Some enterprises were rather straightforward, others were perceived as risky beforehand and were subjected to external evaluation. Yet, it is intriguing to see that several of the projects where the professors were involved, ended in complete fiascos.

One of the envisaged solutions was initiated by Claude Léopold de Genneté.[33] This Lorraine born inventor had studied mathematics with 's Gravesande from 1738 onwards, and it is not unlikely that he witnessed the professor's improvements on Fahrenheit's *waterbuysmolen* at that time. The Dutch governors' interest in Genneté's devices was probably aroused in 1749 when Willem Bentinck van Rhoon – diplomat, advisor of the Stadholder and a member of the Rijnland district water board since 1728 – was on a diplomatic mission in Vienna, where Genneté's inventions were demonstrated. The promise that his designs would turn out both cheaper and better than traditional scoop wheels brought the States of Holland to inviting Genneté to the Netherlands. With the prospect of impoldering the Haarlemmermeer, a patent and a huge sum of money, he was given a lot of land near the old pest house in Leiden to put his inventions to the test (*Fig. 3*). A commission, whose members included the professors Pieter van Musschenbroek – who had moved to Leiden – and Jean Allamand, 's Gravesande's successor as professor of philosophy, would evaluate Genneté's installation upon completion.

Genneté worked on his device continuously. Onlookers were not allowed to come closer than 100 yards.[34] But unfortunately Genneté experienced repeated setbacks. Only after three years, in 1754, could Bentinck

---

32 See Huib Zuidervaart's article, *infra*, pp. 1 ff.

33 The involvement of the Leiden professors in Genneté's enterprise has been described in August C.J. Habets, "Het waterrad van Léopold Genneté", *Spieghel Historiael* 13 (1978), pp. 620–625.

34 John Smeaton, *John Smeaton's Diary of His Journey to the Low Countries 1755*, Leamington Spa, 1938, pp. 32–33.

Fig. 3. Engraving of the hydraulic contrivance Léopold de Gennetté built in Leiden; *Dichtharpoen voor den befaamden werktuig-kunstenaar Gennetté, en zyne befaamde pesthuis-pomp*, [s.l.], c. 1756 (Utrecht University Library, sign. ODJ 1274).

inform the commission that the machine was completed, save for a small leak that still needed to be stopped. Van Musschenbroek and Allamand inspected the – stagnant – installation and made their verdicts.[35] Van Musschenbroek did not have high hopes on the project, even if the leak could be fixed. He judged the installation's principle "entirely wrong and mistaken". Gennetté had attempted to construct an enormous perpetual motion. He had hoped that two mutually connected pumps would maintain each other's motion, but unfortunately this did not happen. An unpublished report records Van Musschenbroek counting Gennetté among the "charlatans and vagabonds", but still he doubted whether this could be true, "for he spoke French fluently". Nevertheless, Gennetté was given the chance to complete his machine – without progress, since he soon made a "clandestine retreat from the country".[36]

Meanwhile the details of the affair were brought into the open. Johan Lulofs – 's Gravesande's successor in the chair of mathematics and astronomy, and from 1754 onwards "inspector general of the rivers" for the

35 The reports have been published in *Nederlandsche Jaerboeken* 16 (1762), pp. 93–166. A summary is in Doorman, op. cit. (n. 28), pp. 310–312.

36 Ibid., p. 310.

States of Holland – took advantage of the situation to impugn Genneté. Some years before, the two had been quarrelling about the mathematics of confluent rivers, and now Lulofs seized the opportunity to publish an anonymous tract on Genneté's installation.[37] In fact the public perception was already settled. At the time Genneté fled the country, another satire, in which both Lulofs and Genneté were ridiculed, found its way to the presses.[38]

Despite being summoned repeatedly, Genneté never returned to Leiden to finish his machine. The Rijnland district water board eventually decided to tear down the installation in 1761, the parts being put up for sale.[39] In the end, Genneté's machine had not met the expectation of replacing traditional technology with cheaper and more efficient tools. But not all hope was lost. Professor Allamand in particular remained confident in striking personalities with original solutions for the Dutch water crisis. On the *perpetuum mobile* builder Bessler, he remarked at a later time that "such madness is incidentally accompanied by geniality. When such characters devote themselves to one particular invention – as Orffyreus apparently did – it is not surprising to see them discover principles that have escaped even the most brilliant minds before".[40] The statement explains why, during the settlement of the Genneté affair, Allamand refused to stand unanimously behind Van Musschenbroek's verdict. Now Allamand set his hopes on François Xavier Trophime d'Arles de Linière, an inventor who had just settled in the Netherlands.[41] The Franche-Comté-born De Linière, had earlier been occupied with hydraulic inventions in France and a later account – written by De Linière himself – described him as a man who achieved perfection in three steps: devising, implementing and, finally, succeeding.[42] But this required investments.

What De Linière had to offer was quite something. He had invented a frictionless pump and, since pumps have the habit of getting stuck in muddy waters, he also had a dredging machine on offer. Both could be driven by a manpowered "new mechanical principle", capable of generating immense powers. Furthermore, the "principle" was universally

---

37 Lulofs, op. cit. (n. 27).

38 *Dichtharpoen voor den befaamden werktuig-kunstenaar Genneté, en zyne befaamde pesthuis-pomp*, [*s.l.*], c. 1756.

39 *Nederlandsche Jaerboeken* 16 (1762), pp. 162–166.

40 Allamand, "Histoire de la vie ...", cit. (n. 8), p. xxvi.

41 Doorman, op. cit. (n. 28), p. 312.

42 François X.T. d'Arles de Linière, *Mémoire à consulter, et consultations, pour la manufacture roïale des nouvelles pompes sans cuirs*, Paris, 1771, p. 4.

applicable.[43] De Linière had an overall solution for the Dutch water management. At the end of 1759 he settled in Voorburg, near The Hague, where he started to unfold his commercial activities.[44] After a demonstration of his devices in The Hague in April 1761, the States of Holland granted him patent for "a new principle applicable to everything that needs moving force, as well as a new type of pumps".[45] The patent for the States General, which followed in 1762, is more revealing about this principle. It was for "a mechanism, to be operated by many workmen and useful for sinking piles, deepening harbours, raising loads, as well as a new type of pump with a solid plunger" (*Pl. VII*).[46] Self-assured, De Linière published a pamphlet emphasizing that his dredger would work 25 to 50 times better than traditional devices.[47]

These testimonies must have been convincing for the States of Holland too. Assisted by their inspector generals of the rivers, Professor Lulofs and Professor Allamand (Van Musschenbroek had died in 1761), the governors freely put a plot of land in Vianen at De Linière's disposal. Its location at the banks of the capricious river Lek made it perfectly suited for De Linière's hydraulic experiments. Initially everything went as planned. De Linière was able to construct a dredging machine with which the water levels could be stabilized. The States of Holland responded with enthusiasm, and asked De Linière to build a second machine for transporting the sediments ashore. But this was never started. De Linière held the opinion that he had already lost enough money on the project, and did a "moonlight flit". Evidently, the governors were anything but pleased with this move. The installations and workshop De Linière had left behind were promptly put up for sale. Meanwhile De Linière continued his work on frictionless pumps in Paris, where he set up a workshop. Here too, things did not go as planned and his business went bankrupt in 1768.[48]

It seems that after this new debacle, the professors once and for all lost their hopes of solving the water problems with mechanical inventions. It is illustrative, for example, that Allamand acquired the old Savery pump

---

[43] F.X.T. d'Arles de Linière, *Memorie of verhandeling over de nieuwe uitvindingen van den heer Darles de Liniere*, c. 1761 (pamphlet deposited at the Archives of the city of Delft, first section, n. 1871).

[44] D'Arles de Linière, *Mémoire à consulter*..., cit. (n. 42), p. 6.

[45] Doorman, op. cit. (n. 28), p. 312.

[46] Ibid., pp. 263–264.

[47] D'Arles de Linière, *Memorie of verhandeling*..., cit. (n. 43).

[48] D'Arles de Linière, *Mémoire à consulter*..., cit. (n. 42), pp. 7–9.

at the auction of Pieter van Musschenbroek's possessions in 1762.[49] Steam technology, with which the professors had already been playing since the 1720s, had grown up and was there to stay. But the professors had some catching up to do. In 1772, Allamand gave a lecture in Rotterdam, outlining "the great advancement our country would make, if steam engines were to be used instead of water mills".[50] At this moment, he also had the privilege of showing the first Newcomen engine in the Netherlands, a model that was built by Edward Nairne in London and that had just arrived by ship. By 1777, no fewer than three steam engines were submitted to trial runs in the city's orphanage.[51] But notwithstanding this, it seems that the professors were lagging behind. The actual implementation of steam power for impoldering the Dutch marshlands in the late eighteenth century was not an academic affair. The initiative came from the Rotterdam watchmaker Steven Hoogendijk. He had been striving for steam-powered water management since the 1750s, and always kept up with the latest developments. It was he who inspired a following generation of engineers finally to impolder the "water wolf", and to institutionalize the Dutch water management of the nineteenth century.[52]

In the previous paragraphs we have sketched how the activities of eighteenth-century Dutch professors of experimental philosophy were not confined to a purely academic setting. A look at their hydraulic ventures in particular shows how the professors' actions were but part of a complex interaction of academic authority, investments, practical expertise and politics. What the professors brought in was their authority on natural philosophical matters. But their actions were not without risk. In an era of growing public interest for experimental philosophy, involving themselves with commercial activities could bring the professors' authoritative status to a corresponding broader audience. In this pursuit, they actively had to prevent losing their grip on the expanding activities of technical innovations. Frequently the professors' involvements were successful, but occasionally they got it all wrong. In situations such as the impoldering of the Haarlemmermeer or dealing with the water crisis, others turned out

[49] De Clercq, "In de schaduw...", cit. (n. 4), pp. 163–165.

[50] The lecture, including a demonstration, is mentioned in *Verhandelingen van het Bataafsch Genootschap der Proefondervindelijke Wijsbegeerte* 1 (1774), p. xliv.

[51] De Clercq, "In de schaduw...", cit. (n. 4), pp. 165–167.

[52] Kornelis van der Pols, "De introductie van de stoommachine in Nederland", in Peter Nijhof (ed.), *Monumenten van bedrijf en techniek*, Zutphen, 1978, pp. 51–61.

to be in a better position for bringing these projects to a good end. Solving these issues required more than just experimental philosophy.

Instrument Cabinets

For the material heritage associated with the professors' hydraulic pursuits, these observations have significant consequences. If the dissemination of experimental philosophy to the public was not a one-way process, we cannot hold the material heritage that constituted the "canonisation" of experimental philosophy to be a coherent set of mere demonstration devices. Although a vast quantity of these instruments, present in eighteenth-century cabinet and textbook culture, did illustrate long-established laws and principles, there remains an interesting portion of instruments and models that are unique in their representation of complex, experimental enterprises in which both academy and society had their share.

But where are these instruments and models now? Contemporary documents provide a good start. We have, for example, the catalogue of the private cabinet of professor Allamand, which was auctioned in 1788. Of interest is the announcement on its title-page, "noteworthy are the different models of Dutch wind- and water mills, pile drivers, etc.".[53] Indeed, while the "classical" physical instruments, with reference to the well-known textbooks, are abundant, there is a separate section of "machine models". Here we find mud dredgers, water mills and other mechanical contrivances, occasionally accompanied by references to contemporary inventors and enterprises. Obviously the working principle's soundness determined the survival chances of certain objects: it is a pity that the "model of a water pump, after the invention of Geneté", listed among Allamand's possessions, did not survive.[54] Others, however, apparently did. There is a model of a mud dredger, originating from Allamand's cabinet, preserved in Museum Boerhaave (*Fig. 4*). Its background becomes a lot clearer when we take into account that De Linière's dredging machine was repeatedly compared with a dredger "that is used to keep the harbour of Toulon deep and navigable".[55] The preserved model shows a very close

[53] *Catalogus van eene aanzienlyke verzameling van allerlije instrumenten*, Leiden, 1788.
[54] Ibid., p. 16.
[55] D'Arles de Linière, *Memorie of verhandeling*…, cit. (n. 43), pp. 2–6. De Linière refers to a dredger described in Bernard F. de Belidor, *Architecture hydraulique*, Paris, 1737.

Fig. 4. Model of a dredger, 1750–1775 (Museum Boerhaave, inv. no. V09696).

correspondence to this Toulon dredger. It is one instance of an object that loses a lot of its meaning if we attempt to interpret it in a solely academic context.

To this we can add several machine models that 's Gravesande added to the Leiden Cabinet of Physics. In 1742, the year the professor died, this contained an enormous, functional polder mill model (*Pl. IV*). Among the other additions was a water-lifting pump after the professor's design (*Fig. 5*), and a model of a scoop wheel (*Pl. VIII*). Close to nothing is known about the backgrounds of these objects but, given the professor's notable hydraulic involvements, it is unlikely that these artefacts served as mere demonstration instruments. It is more likely that they served as experimental devices, representations of enterprises, or even patent models.

Yet, occasionally it is possible to reconstruct the complex meanings an artefact can hold, by combining the information available from written sources, catalogues and the artefact itself. In Pieter van Musschenbroek's 1762 *Introductio ad philosophiam naturalem* we find a depiction and brief description of a peculiar-looking pump. Van Musschenbroek notes about the device that "it was the talented craftsman Jan Paauw who presented me a fine model of this instrument".[56] So far, this appears to be an instance of a scholar disseminating mechanical principles by presenting a machine design he had at his disposal. But a look beyond the traditional sources is more revealing. In the auction catalogue of the cabinet of Amsterdam

[56] P. van Musschenbroek, *Introductio ad philosophiam naturalem*, Leiden, 1762, pp. 873–874.

Fig. 5. Water-lifting pump after an invention by 's Gravesande, from the professor's private cabinet. Jan van Musschenbroek, 1742 (Museum Boerhaave, inv. no. V09698).

mayor Gerard A. Hasselaer, who was also involved in De Linière's enterprises, we find a pump model that, due to its survival in the collections of Utrecht University (*Fig. 6*), proves to be identical to Van Musschenbroek's description.[57] But what we learn from the auction catalogue too is that this model represents a "frictionless pump" after the design of De Linière.

[57] *Catalogue d'une très-belle collection d'instruments physiques*, Utrecht, 1776, p. 7. Also see Tiemen Cocquyt, "The Hasselaer Auction of 1776: The Transmission of Scientific Instruments from the Public to the Academic Sphere", *Rittenhouse* 22 (2009), pp. 70–89.

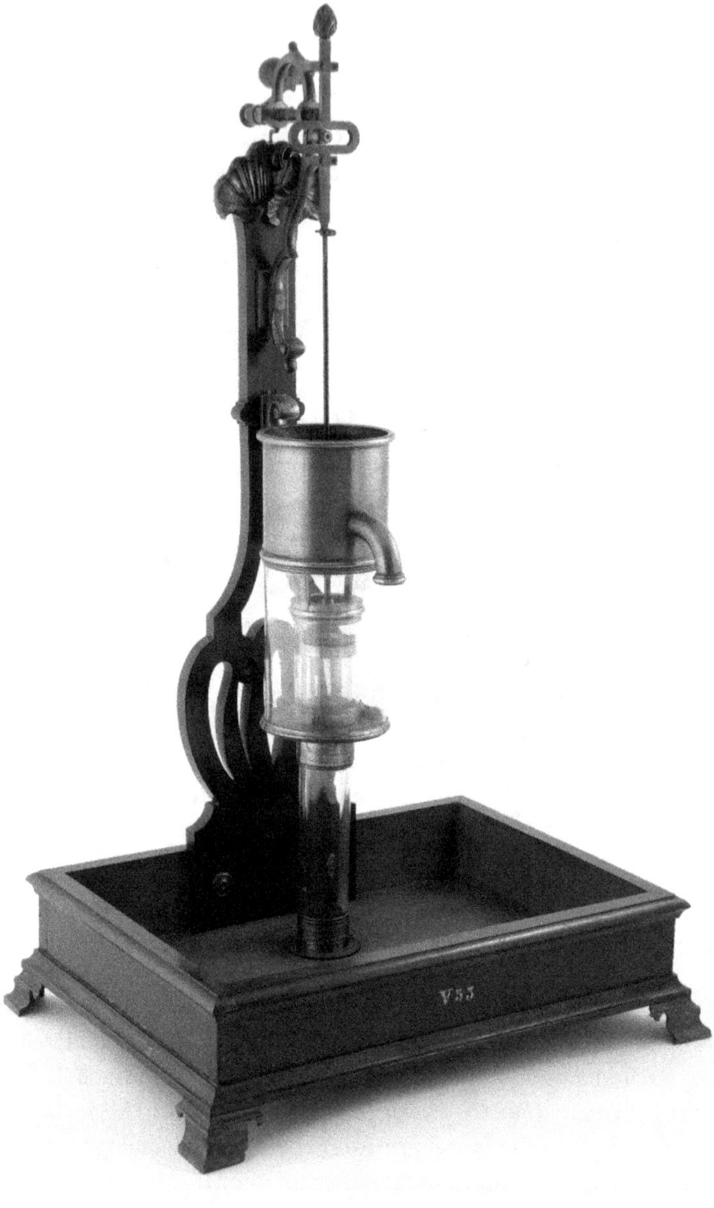

Fig. 6. Model of De Linière's "frictionless pump" by Jan Paauw, c. 1760 (Utrecht University Museum, inv. no. V53).

It turns out to be the very design Van Musschenbroek and his colleagues were so hopeful of in the early 1760s, when they brought their academic expertise into action in attempting to solve the Dutch water crisis.

In this respect, the request by Allamand's successor Christiaan H. Damen to the University Board, to get rid of those instruments which are of use "neither for the study of the history of machines, nor for the instruction of physics" is intriguing. It seems that a clearing-out was desired. There definitely was a place for the, perhaps slightly outdated but still valued, historic cabinet testifying to the great heights Leiden experimental philosophy had reached. But the history it bore witness to had to meet certain criteria. Instruments that were useful for demonstration, or machine models illustrating sound mathematical principles were kept. But artefacts associated with more dubious experimental enterprises were not always passed down to present times. And even if they were, the background they testify to often evaporated in the transformation of science-in-the-making to canonical experimental philosophy. Instrument cabinets were affected by the process in which experimental philosophy was shaped, and that is something to keep in mind when assessing them as relics of the past.

THE CABINET OF PHYSICS AT RIDDARHUSET IN STOCKHOLM IN THE EIGHTEENTH CENTURY

Inga Elmqvist Söderlund

Demonstrations of Experimental Physics at Riddarhuset

Riddarhuset is situated in a good central location in the Old Town of Stockholm, and is the House of Nobility, a Swedish equivalent to the British House of Lords. Sweden was ruled by the so called "Ståndsriksdagen", or "Riksdag of the Estates". It was a Diet made up of the four Estates, which were the Nobility, the Clergy, the Burghers and the Peasants. Riddarhuset was built in order to house political meetings and for the administration of the Nobility. The imposing building was erected in the middle of the seventeenth century, at a time when Sweden was relishing the successes of its expansive foreign policies. The building is an obvious statement of the importance that the Nobility assigned to itself in political terms (*Fig. 1*).

In eighteenth-century Sweden the natural sciences were becoming increasingly popular. Riddarhuset became an important site in this development, housing lectures on varying subjects and also musical concerts. With the establishment of a cabinet of physics, it also became the main meeting place for those inclined towards natural philosophy. The possibility of establishing such a cabinet at Riddarhuset originated in events in the late 1720s.[1] Sebastian Tham, a merchant from Gothenburg, donated financial funds in 1727. The idea to use Riddarhuset for teaching and lectures was not novel, as in the seventeenth century the premises had housed a school for young noblemen, the "*Collegium Illustre*", although this seems to have functioned only briefly. The Thamian bequest was used throughout the eighteenth century to teach natural philosophy, in

[1] The main published works on the cabinet are: Sten Lindroth, *Kungl. Svenska Vetenskapsakademien Historia 1739–1818*, Stockholm, 1967, 2 vols.; Svante Lindqvist, *Technology on trial*, Uppsala, 1984; Henrik Schück, "Ansatser till ett universitet i Stockholm före 1800-talet", *St Eriks årsbok* (1942), pp. 7–68; Anna Beckman, "Två svenska experimentalfysiker på 1700-talet: Mårten Triewald och Nils Wallerius", *Lychnos* (1967–1968), pp. 186–214; Karl Grandin, "Den Thamska donationen och de Thamska föreläsningarna", *Documenta, Berättelse över Kungl. Vetenskapsakademiens verksamhet* 69 (1999), pp. 63–65.

Fig. 1. Riddarhuset from lake Mälaren; engraving from Erik Dahlberg, *Suecia Antiqua et Hodierna*, Paris and Stockholm, 1667–1715, vol. 1 (Photograph: The National Library of Sweden).

the form of lectures and publications. The funds still remain but are today used by the Royal Swedish Academy of Sciences to invite particularly distinguished lecturers.

This paper will focus on the lectures of experimental physics and collections of scientific instruments held at Riddarhuset from 1728 onwards, at first by Mårten Triewald, then by other scholars within the framework of the Royal Swedish Academy of Sciences. It will track the present whereabouts of the instruments as far as they are preserved and known. It will also briefly describe the other cabinets of physics that were set up in Sweden in the eighteenth century, in Stockholm, Lund and Uppsala.

In 1728 the first lectures in experimental physics were presented in facilities on the ground floor at Riddarhuset (*Rikets ständers secrete utskotts salen*). These rooms would house important political meetings for about a hundred people. It provided a good start, but seems not to have been a satisfactory solution, as a few years later, in 1734, Mårten Triewald made efforts to have the unused eastern part of the second floor refurbished into a "*Schola illustris*" with space for necessary instruments, machines and a lecture hall. This was granted and Triewald was able to start his renewed public activities in the "*Auditorium Illustre*" in 1739. The rooms have unfortunately been refurbished, but a drawing gives an impression of the layout (*Fig.* 2). A round theatre with tiers of seats connected to stairs and a stage in the middle provided the main space for lectures. Outside the theatre was an iron barrier, behind which was a repository for "machines, models and instruments". The rooms adjacent to the theatre comprised a vestibule and repository for larger instruments, and a "*Laboratorium Mechanicum*" or room for experiments. There was also a particular repository for models of fortification and room for drawing for the young men of the Royal "*Lifgardet*", the Life Guards of the Army. This suggests a cooperation with education for military purposes.

In the same year as Triewald started his activities in the *Auditorium illustre*, the Academy of Sciences was founded at Riddarhuset, with Triewald as one of its founders, and main proponent, proposing the Royal

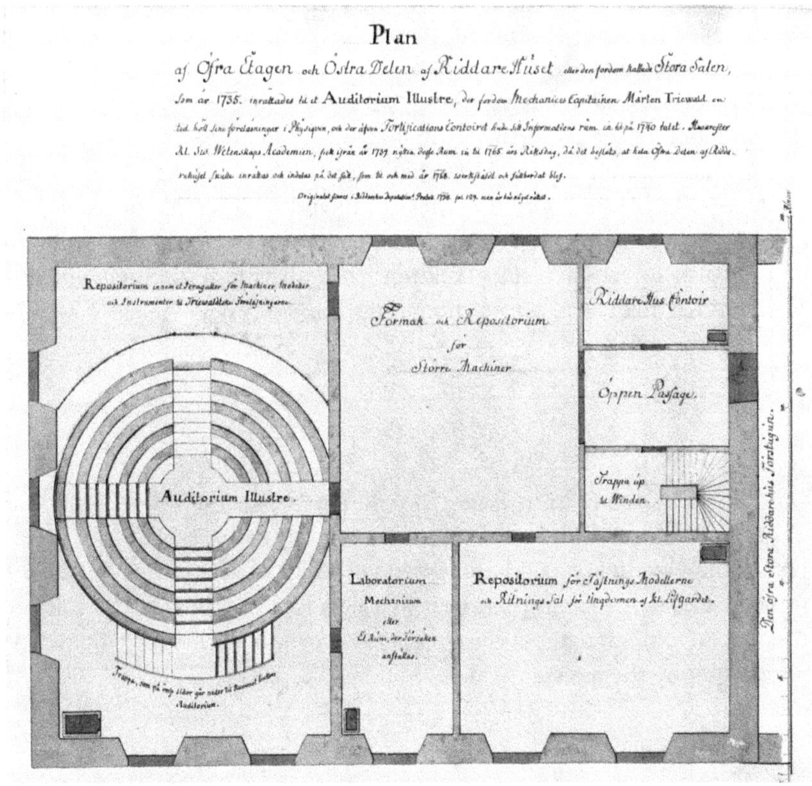

Fig. 2. The "*Auditorium Illustre*" and repository of instruments at Riddarhuset, as used from 1737 to 1768; floor-plan, Tilas collection, t. 1 s. 102 (Photograph: The National Library of Sweden).

Society as a model. The Academy was granted the use of the rooms for 25 years, but thereafter the Academy continued to host lectures and official events at Riddarhuset. Physics was not the only subject for lectures at Riddarhuset. For example, Carl Linnaeus, another founding member of the Academy, held lectures on botany, zoology and mineralogy. They were so popular that Triewald's room was too small.[2]

The main audience was probably initially a genteel, male clientele and merchants; later, as the lecture series became more established, it provided useful training for fortification officers and architects. The first advertisement of Triewald's lectures addressed the audience as "honoured Gentlemen". The list of subscribers for the published lectures contains the names of several well-known and wealthy men. The printed lectures also contain illustrations of instruments, each plate with the subscriber's coat of arms. This is an implicit reference to the great hall of Riddarhuset, which holds the coat of arms of the nobility. The inclusion of the coat of arm situates experimental physics within the nobility.

We cannot exclude female interest, but it is reasonable to assume that the audience of the lectures was male and the main target group for the books male as well. In the latter half of the century Johan Carl Wilcke made a point of the appropriateness of including a female audience. He advocated the usefulness and pleasure to be derived from experimental physics also for women and children. He particularly pointed out that Madame Châtelet was a role model, not only learned, but also writing and instructing others.[3]

Mårten Triewald

On 8 January 1728 Mårten Triewald held his first lecture of experimental physics at Riddarhuset. It was the starting point in a series of 28 lectures. It was followed by a second and extended series of 30 lectures starting on the 15 October the same year. In 1730 a new series of lectures was advertised, now extended to 32 lectures; they were to be held by Triewald's permission by the assistant Daniel Menlös.[4]

---

[2] Carl Linnaeus, Letter to Carl Gustaf Tessin, 22 April 1740, in Ewald Ährling (ed.), *Svenska arbeten*, Stockholm, 1878–1880, 2 vols.; v. 1, pp. 1–3.

[3] Johan Carl Wilcke, *Tal, Om naturkunnighetens nytta, behaglighet och Uphielpande*, Stockholm, 1762, p. 18.

[4] Schück, op. cit. (n. 1), pp. 7–68.

In a portrait of Triewald attributed to Georg Engelhard Schröder, preserved at the Royal Swedish Academy of Sciences, he is shown in half-figure, pointing at a bird within a bell jar (*Pl. IX*). This gesture is at once a reference to his teaching and to his research in pneumatics. He is best known for his contributions to the developments of the steam engine as well as his publications on the diving bell: *The Art of Living under Water*. When he performed his first lectures he had just come back from England, where he had stayed for eleven years. In London he had attended lectures on natural philosophy by John Theophilus Desaguliers, had seen how experiments were performed and demonstration instruments set in front of an audience. He was impressed by the Royal Society, of which he later became a fellow. Newton, whom he claimed to have met, became the authority to whom he often referred. Triewald started to teach in rented rooms in Newcastle and he lectured on astronomy and natural philosophy in Edinburgh in 1724 and 1725.[5] He perceived the widespread interest in natural philosophy in England. He wrote that in several towns in the country, philosophy, by which he probably meant experimental natural philosophy, was taught, and lecturers were competing with each other. He bought a whole set of instruments, so it is fair to assume that he was teaching on a more regular basis. He claimed to have bought the set on his own account, with no financial help from home, and brought it back to Sweden, aiming to introduce the new fashion to his home country.[6]

A printed *Notification* advertises the lectures on mechanics, hydrostatics, aerometry, and optics on 16 pages, signed by Triewald on 2 September 1728. It is a description of what each lecture would treat. Experiments and instruments play a key role in the text, apparently thought to be an appropriate means to promote the series. From the text it is obvious that the first lectures in the beginning of the year went very well, but that Triewald was marketing the event to gain a larger audience. The lectures were to be held every Tuesday, Thursday and Saturday at five o'clock sharp, to be paid in advance, at a shop in the square outside Riddarhuset, and the receipt was to be presented at the entrance of the lecture room.

---

5 Lindqvist, op. cit. (n. 1), pp. 197 ff.; John A. Cable, "Early Scottish Science: The Vocational Provision", *Annals of Science* 30 (1973), pp. 179–199; Svante Lindqvist, "Social and Cultural Factors in Technology Transfer", in Kristine Bruland (ed.), *Technology Transfer and Scandinavian Industrialisation*, New York & Oxford, 1991, pp. 15–36.

6 Mårten Triewald, *Mårten Triewalds år 1728 och 1729 håldne föreläsningar*, Stockholm, 1735–36, 2 vols.

The auditor could simultaneously sign up and pay for the published version of the lectures, and the illustrations.[7]

A written account of the lectures was published in two tomes printed in 1735–1736. Each is richly illustrated, in all with 64 plates. This edition was reprinted in 1758. Unfortunately the text treats only half the lectures as the planned parts three and four were never published. The first lecture started out on very small bodies and the point of departure is Newtonian physics. Some instruments and experiments are mentioned: microscopes, magnets, and a vacuum experiment with mercury. The first plate shows a glass globe friction machine very similar to Francis Hauksbee's (*Fig. 3*). Triewald wrote that before using the machine, grey paper or dry wool was rubbed on a glass staff, then a gold-leaf was thrown into the air, whereby he could show how it could be moved in the air by the staff. Then he turned to the machine, which had a hollow glass bulb, pierced by an iron rod resting on supports outside the bulb. Within the bulb was a round board to which were attached strings made of wool. The bulb was turned by a crank connected to the bulb by a string. Suspended above the glass bulb was a bow from which hung strings of wool. As soon as the experiment was begun, the outer strings would stiffen and point towards the middle, the inner strings would straighten and point outwards like the spokes of a wagon wheel.[8]

Triewald's style of teaching in vernacular appealed to the audience, but what pleased many, also made him enemies. He referred to this in the publication. He wrote that they did not like the lectures to be in Swedish. Triewald complained that a few were not really interested, but were there simply to pass the time. The criticism may also have been aimed at the great number of experiments, such that there was no time left to explain the theory. For example, the *Notification* states that lecture 15 contained 50 experiments with the air pump.[9] It is possible that he used different senses to heighten the attention of his auditors, and that this was associated with street-showmanship. It seems a lot was happening on the stage. He wrote on the mixing of liqueurs, making drinks with liquids in layers of different colours, and he referred to smell (smoke) and also a loud bang when the thermometer broke. The eighth lecture treated the living forces of humans and other animals. This was demonstrated by men on

[7] Triewald, *Notification, om trettijo publique föreläsningar*, Stockholm, 1728.
[8] Triewald, *Mårten*..., cit. (n. 6), v. 1, pp. 23–25.
[9] Triewald, *Notification*..., cit. (n. 7), un-paginated [f. 6v].

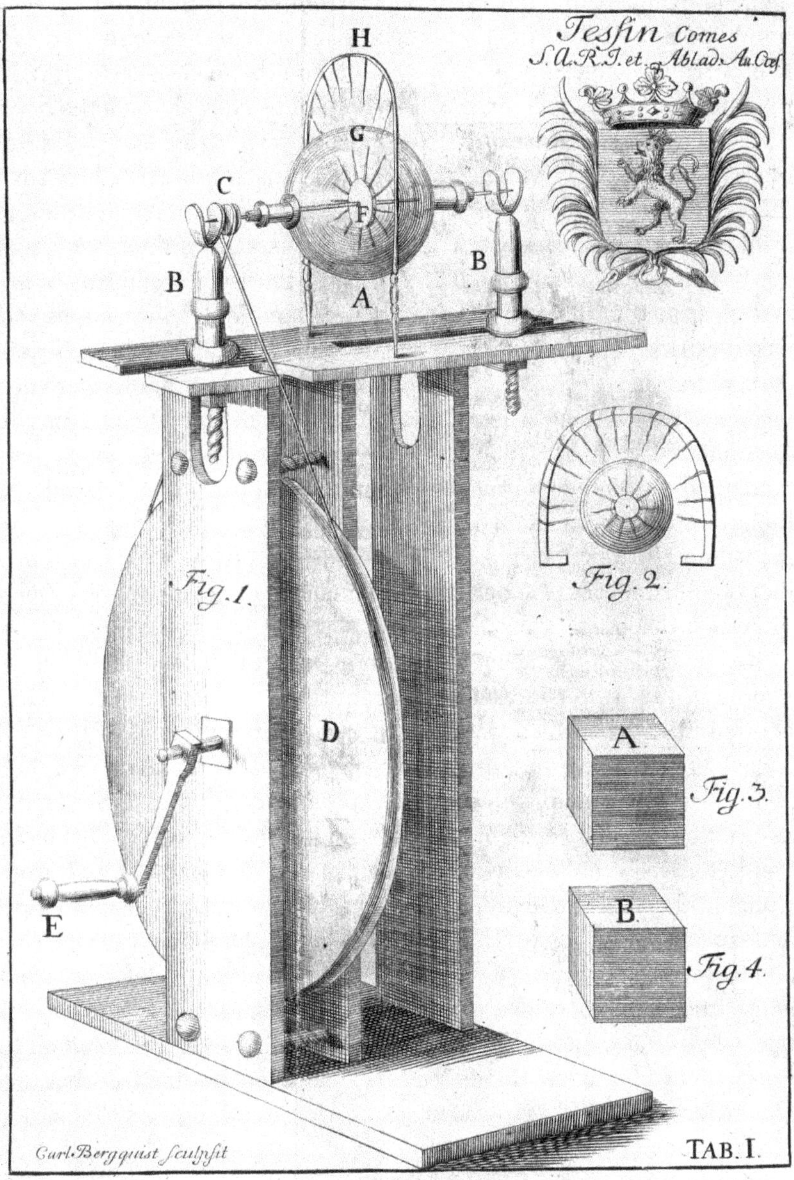

Fig. 3. Glass globe friction machine with the coat of arms of Count Tessin; engraving by Carl Bergquist, in Mårten Triewald's published lectures, vol. 1, 1735 (Photograph: The National Library of Sweden).

the stage, one being referred to as particularly strong. In one of the experiments a man used the force of a sledgehammer to hammer on a piece of iron laid on the breast of another man lying flat on the ground. Triewald wrote that the larger the piece of iron laid on the man's chest is, the less the force of the sledgehammer will be felt. Here Triewald took the opportunity to reveal that travelling strongmen, performing such an act in front of an audience, may not be extraordinarily resistant to heavy blows with the sledgehammer after all, the large piece of iron protecting the man from the force of the blow.[10] There is not necessarily a direct correspondence between what was said and done in the lecture and was written and illustrated in the printed version. For example the illustrations show naked muscular men, covered only with a bit of cloth, while performing demonstrations (*Fig. 4*). It is improbable that demonstrations were done in such scant clothing. These illustrations were in fact copied from John Theophilus Desaguliers' book *A course of experimental philosophy*, even if adapted to the new format and context.[11] It is likely that Triewald had the artists copy instruments and experiments from different sources and that some instruments were drawn from life. Otherwise it can be assumed that the illustrations give quite an accurate idea of the instrumentation at Triewald's disposal.

The particularity of keeping and showing a cabinet of physics in the middle of the winter in a northern country is evident from an account of a curious event which was published in the transactions of the Royal Society. Triewald wrote that coming into the hall with some visitors he feared that the glass for showing the experiment with the "Cartesian devils" would be in danger of breaking, should the water freeze. This instrument consists of hollow divers (figures) placed in a jar filled with water, which at the pressure on a membrane on the surface, are made to move, and sink to the bottom of the glass. He took the jar down from the shelf, and was well pleased to see the water was in a fluid state. Before he would empty the glass, he wanted to show the experiment to his visitors, but in that very instant, and in the space of a second of time, he found all the water changed into ice. He offered no explanation, but invited "that ingenious Gentleman Dr. *Desaguliers*" to provide one.[12] Besides this curious

---

[10] Triewald, *Mårten*..., cit. (n. 6), v. 1, pp. 12–14, 31–38 and 285–327.

[11] John Theophilus Desaguliers, *A course of experimental philosophy*, London, 1734–1744, 2 vols.; v. 1, pl. 19.

[12] Mårten Triewald, "A letter from Mr. Triewald ... to Sir Hans Sloane ... Relating to an Extraordinary Instance of the Almost Instantaneous Freezing of Water", *Transactions of the Royal society* 37 (1731–32), pp. 79–81: 79–80.

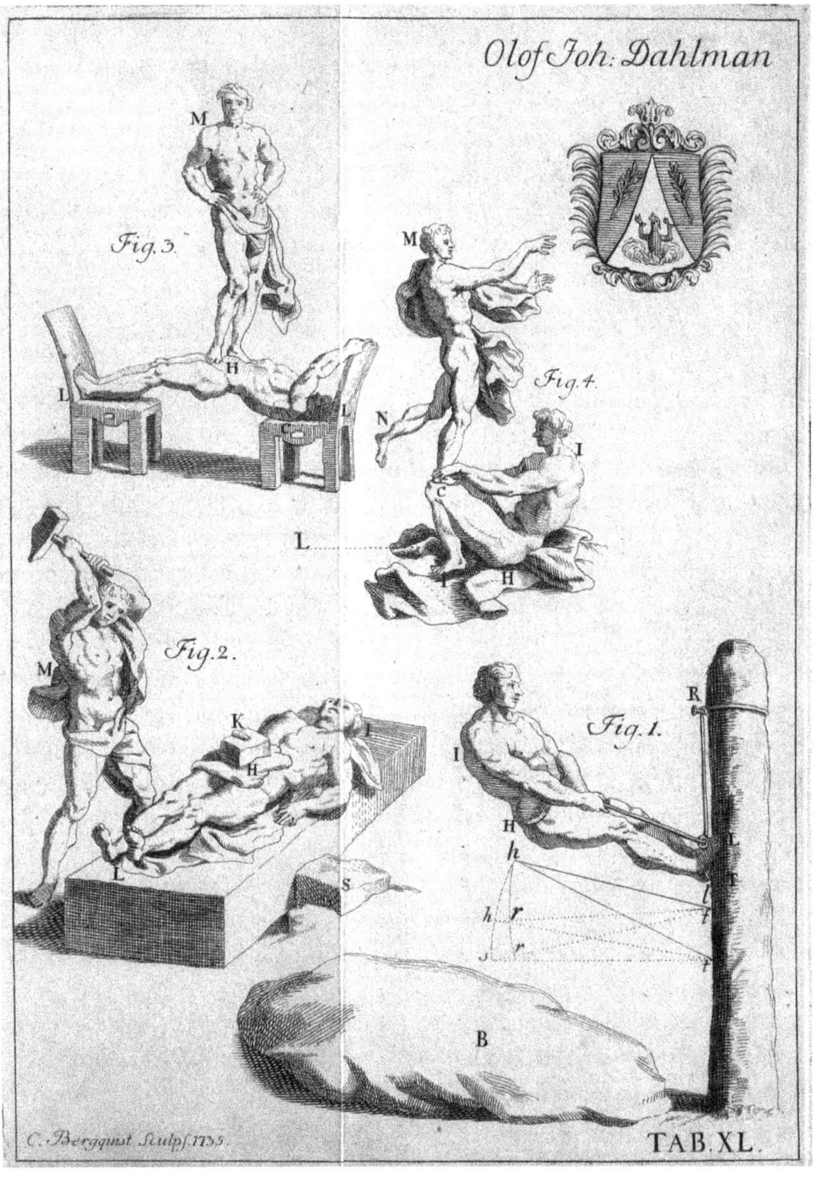

Fig. 4. Men demonstrating the forces of the human body; engraving by Carl Bergquist, in Mårten Triewald's published lectures, vol. 1, 1735 (Photograph: the National Library of Sweden).

instant freezing we can wonder at the fact that the rooms were not heated to protect the artefacts or when visitors were expected.

In 1732 Triewald sold his collection of 327 instruments to his assistant Daniel Menlös, who brought it to Lund University (see below). Triewald acquired a new collection, but its content and destiny is not known and neither is how regularly he lectured. In the commemorative speech after his death, presented at Riddarhuset, it was indicated that the instruments were still preserved at the Academy. His instruments were probably displayed at Riddarhuset in 1747, and possibly still in 1780.[13]

The Cabinet of the Royal Swedish Academy of Sciences

In 1746 it was decided that that the permanent secretary of the Academy of Sciences, Pehr Elvius would hold the Thamian lectures at Riddarhuset. This decision was in part founded on financial grounds as it allowed the secretarial post of the Academy to be salaried for the first time. The secretary was to hold lectures every Wednesday at three o'clock. In the commemorative speech after Elvius's death, it was pointed out that his prime model for teaching was the activities at the Florentine academy, by which was probably meant the seventeenth-century *Accademia del Cimento*. As he did not travel widely, the inspiration for teaching must have come from experiences in Stockholm and Uppsala, and not least from books or from someone who had travelled abroad. Possibly his teacher and well travelled cousin Anders Celsius had told him about the wonders and collections of instruments in Italy. In all Elvius performed fourteen lectures. The audience stopped coming and Elvius's lecturing duties were removed. He still kept the salary though, and his new task was to write and publish an article on the history of the sciences in the transactions of the Academy four times a year. His contemporaries appreciated his skills in writing more than his lecturing.[14]

When Elvius died in 1748, he was succeeded by Pehr Wargentin. Performing lectures was not something Wargentin prioritised, although its importance was stated repeatedly. Wargentin wrote: "To read for the public with quality would take all your time, particularly if you want to do experiments as is stated in the directions [of the will of Sebastian Tham]

---

13 Lars Laurell, *Åminnelse-tal öfver... Mårten Triewald*, Stockholm, 1748, p. 22; Carl Wilhelm Oseen, *Johan Carl Wilcke experimental-fysiker*, Uppsala, 1939, p. 307.

14 Olof Dalin, *Åminnelse-tal öfver... Pehr Elvius*, Stockholm, 1750, pp. 20 and 25.

and also what the audience desires".[15] Wargentin emphasized the audience's wish for experiments and instruments, and the absence of an audience was blamed on the lack of appropriate instruments. It is very possible that he was not a very captivating reader and soon another solution was sought. The complaint of lack of instruments is somewhat contradicted by other evidence that instruments were available. For example, the Academy had acquired an air pump and an electricity machine in the 1740s, and in 1755 and 1757 the accounts indicate payments to transport an electrical machine from Riddarhuset to the observatory, where Wargentin lived and performed his research.[16] At the observatory he was neighbour to the instrument maker Daniel Ekström, who also contributed to the instrumentation by making the large electrical machine for the Academy, but his untimely death in 1755 put an end to the cooperation between the two.[17]

Johan Carl Wilcke

In 1759 Wargentin managed to persuade the other members of the Academy to employ the young and probably charismatic Johan Carl Wilcke. The financial situation had improved and allowed the creation of the "Thamian" lecturing post. Wilcke was to teach twice a week for two months in spring and autumn. This arrangement allowed him time also to develop as a researcher, particularly within electricity, developing an electrophorus in 1762, and to work on "specific heat".[18] The salary was however on the meagre side and, in order to make a decent living, he took a side job as tutor to the son in the family Höpken. Wilcke managed to revitalize experimental physics in Stockholm, holding a great number of lectures, and under his supervision the collection of instruments grew substantially.

Wilcke had studied at Uppsala where he had witnessed the use of instruments in Samuel Klingenstierna's demonstrations. Klingenstierna had acquired a full set of experimental physics instruments from London and translated Pieter van Musschenbroek's *Elementa Physicae* into

---

15 Nils Nordenmark, *Pehr Wilhelm Wargentin*, Uppsala, 1939, p. 154.

16 Royal Swedish Academy of Sciences, Center for History of Science (RSAS-CfVH), Stockholm: Accounts, 1755, no. 38, and 1757, no. 63.

17 Olov Amelin, *Medaljens baksida*, Uppsala, 1999, pp. 74–76.

18 Oseen, op. cit. (n. 13), pp. 68–276.

Swedish. In the foreword Klingenstierna wrote on the usefulness of instruments in teaching, about how difficult it is to obtain knowledge about nature, and that the "powers are hidden to our senses, as their function is invisible, and must be explained by particular accounts and made visible by the aid of *instruments*".[19] If Klingenstierna managed to drive home the usefulness of demonstrational instruments in teaching, Wilcke gained further opportunity to learn by travelling abroad. In 1751 to 1757 Wilcke travelled to Germany, where he studied at Rostock and Berlin, undoubtedly becoming acquainted with teaching and collections, making friends with, among others, Franz Aepinus. In his texts, Wilcke referred to many contemporary European colleagues, as well as printed works, aspiring to make Stockholm more up to date in comparison to what was being done on the Continent and in England.

Some of Wilcke's lectures and speeches were published. In 1761 he held a lecture in which he argued for the usefulness of the acquisition of instruments for the demonstration of experimental physics. He emphasized that teaching of experimental physics must be pleasant and needs instruments in order to attract an audience.[20] At his death in 1796 the Academy acquired Wilcke's private collection of 287 items, of which some had been used in his lectures at Riddarhuset, but it is not easy to know exactly which instruments were demonstrated when. There was a variety of instruments relating to pneumatics, magnetism, optics, acoustics and, not least, electricity. There were several electrical machines along with isolator tables, well suited for demonstration, and we can gather that electricity was a key element on the programme. More than 70 artefacts related to Wilcke are identified as part of the Academy's collection.[21]

The accounts of the Academy indicate that increased amounts were spent on the lectures of experimental physics after the employment of Wilcke. Several instruments were paid for, but it is not always clear if they were intended for the cabinet of physics, the observatory or for something else. In 1759 Anders Smahl, the janitor at the observatory, was granted a salary increase, because he was to assist with the lectures. In 1760, tips were given to the instrument maker Steinholz's boys to bring an air pump to Riddarhuset and three days later the turner was paid for glass tubes, bladders, gold leaves and a particular oil for the machine. In 1761 the

---

[19] Samuel Klingenstierna, "Foreword", in Pieter van Musschenbroek, *Inledning til naturkunnigheten*, Stockholm, 1747, un-paginated [f. 2v].

[20] Wilcke, op. cit. (n. 3), exp. p. 32.

[21] Gunnar Pipping, *The Chamber of Physics*, Stockholm, 1977, pp. 84 ff.

machines for the lectures were repaired.[22] Unfortunately the air pump mentioned here seems to have been lost, as it was replaced in the 1780s or 1790s with one made by Johann Heinrich Hurter of London (*Fig. 5*). This artefact was given the no. 1 in the Academy's inventory of 1798, whereby it was often wrongly referred to as the first instrument of the collection. The first proper demonstration instrument had been given to the Academy by its member Jacob Faggot, probably in the early 1740s.[23] In this case it seems it was more the function of the object than the provenance that was relevant in the inventory. This shows how important it is to look at the artefacts critically and evaluate whether provenance information in inventories is credible, and whether items have been replaced or not.

Some identified artefacts are very small and would not be effective for a large audience seated far from the lecturer. For example two small electric dolls, Harlequin and Colombine, fashionably clothed and brightly coloured, are only 11 centimetres tall. Their dancing floor is a brass disk, the insulating foot now missing (*Fig. 6*). To be able to enjoy their jerking electric dance, the beholder would need to be rather close to the performance.

In 1764 Wilcke was reimbursed for moving instruments to the Academy's house at Helgeandsholmen (The Old Bank), after which the collection had an itinerant existence for some years. In 1767 Wilcke was granted an allowance to rent other rooms for his lectures. In 1771 the instruments were moved to the Academy's new headquarters at Greve Pehr's Hus, and the rooms were decorated and shelves installed.[24] Soon the Academy was to move again, as a new building was bought in 1779 at Stora Nygatan. In 1780 Wilcke affirmed that the Academy was in possession of a good collection, but there was no appropriate place for it.[25] Eventually rooms were set up on the third floor at Stora Nygatan. The instruments were lined against the walls and in the middle of the floor stood globes. However, at the end of his life, Wilcke complained it was nothing but storage: there was hardly any space for experiments. At this point he had pursued his career within the Academy and acquired the shared post of permanent secretary; other duties had taken him away from physics. Even if Wilcke did not teach any

22 RSAS-CfVH: Accounts, 1759 no. 172, 8 May and 31 October; 1760, nos. 229, 245–246, sec. II no. 65; 1761, no. 112, 161.

23 Pipping, op. cit. (n. 21), p. 177.

24 RSAS-CfVH: Accounts, 1764, no. 66; 1771, no. 265; 1772, no. 100, 146, Minutes 18 February 1767.

25 Oseen, op. cit. (n. 13), pp. 307 ff.

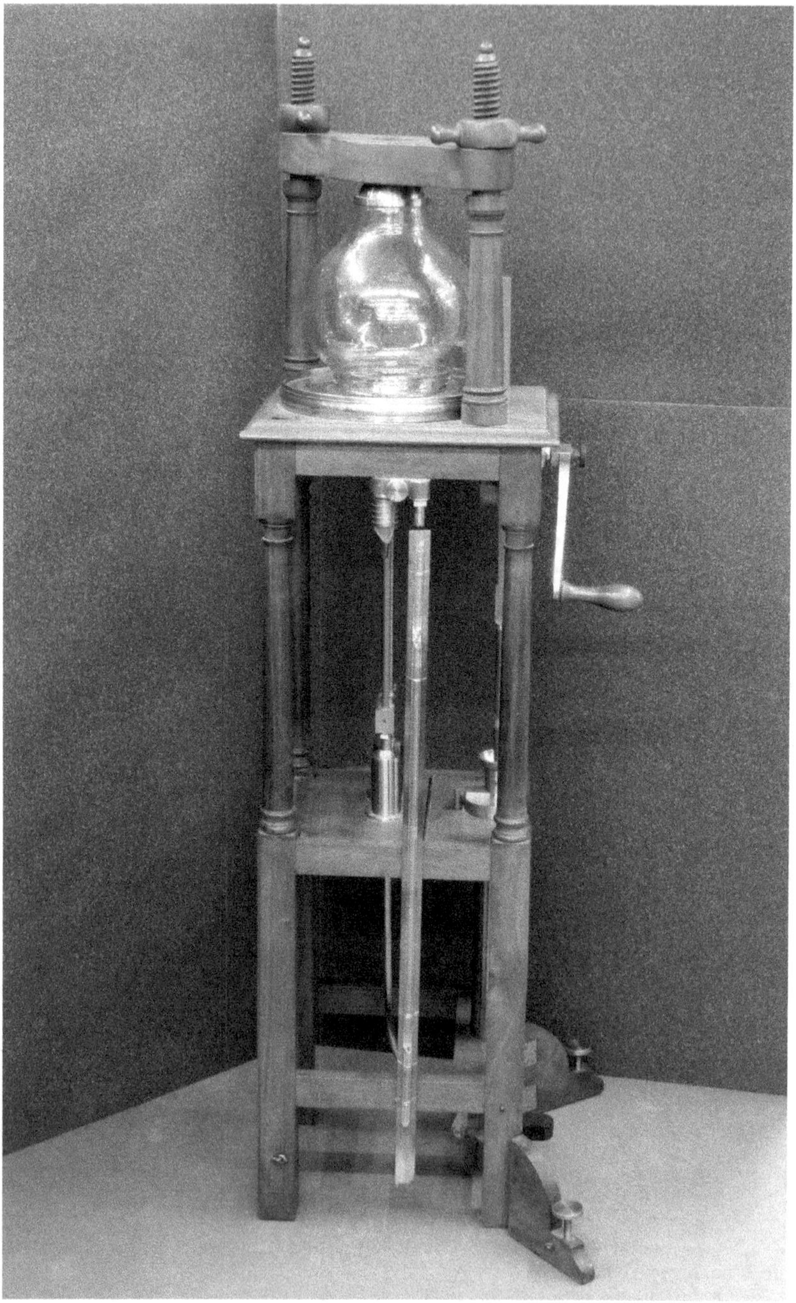

Fig. 5. Air pump by Johann Heinrich Hurter, London, ca. 1780–1790; KVA 1 (Photograph: Center for History of Science, Royal Swedish Academy of Sciences).

Fig. 6. Electrical dolls: Harlequin and Colombine, and brass disc dancing floor, from the Wilcke collection; KVA 1521 (Photograph: Center for History of Science, Royal Swedish Academy of Sciences).

more, the collection was still used: for example in 1794 Baron von Gedda borrowed instruments to teach the young King at the Palace.[26]

In the nineteenth century the Academy moved premises again, to Drottninggatan, and in the early twentieth century it moved to its present location in Frescati. The instruments were moved as well and also dispersed, discarded and neglected, as they fell out of fashion. In the latest move, the

[26] Lindroth, op. cit. (n. 1), v. 2, pp. 364 ff.

physicist Wilhelm Carlheim Gyllenskiöld recognised the historical value of the collections and started to trace the history of the objects. Further work with cataloguing was undertaken particularly by Gunnar Pipping, but there is still a great potential to discover new information in the extensive archives, and not least from the instruments themselves.

Overview: Collections of Scientific Instruments in Eighteenth-Century Sweden

There were three main cabinets of experimental philosophy in Sweden in the eighteenth century. The first, the one which is the focus of this article, was at Riddarhuset in Stockholm and the other two were kept at the universities in Lund and Uppsala. There were also other collections which provide a frame of reference, from which the lecturers at Riddarhuset at times could borrow, trade and exchange artefacts.

The research and teaching performed at the universities of Lund and Uppsala required instruments, but many were the private property of the individual professors. As has been mentioned above, in 1732 the University of Lund acquired Mårten Triewald's first collection used at Riddarhuset. Triewald's assistant Daniel Menlös developed as a lecturer in his own right. In 1732 he was appointed professor of mathematics at Lund University on the condition that he would at his own expense acquire Mårten Triewald's collection of instruments and donate it to the University. In all, Menlös listed in an inventory 327 items to be donated. The collection was shipped in thirty boxes from Stockholm to Lund. The collection was placed at Kungshuset in Lundagård, where an auditorium was constructed. Today about seventy of Triewald's original items are preserved; they belong to the department of physics, Lund University, but are deposited in Malmö Tekniska Museum (*Pl. IX* and *Fig. 7*). In the inventory from 1732 the artefacts are described according to material, design and what the experiment demonstrates. References are made to pages and plates in the books by Willem Jacob 's Gravesande, Francis Hauksbee, Christian Wolff, Christian Gottlieb Hertel, and Jacob Leupold. In only a few cases are the names of the instrument-makers specified (Hauksbee, Barclay and one Swedish instrument maker, Johan Herbst in Stockholm). The list also specifies an air pump made by Otto von Guericke, which, except for a few exchanged parts from the eighteenth century, appears to be the original. It seems that the instrument can be traced through different owners from

Fig. 7. Model of the eye, belonged to Mårten Triewald; Malmö Tekniska Museum, T 197 (Photograph by courtesy of Skotte Mårtensson, the Faculty of Engineering at Lund University).

1663.[27] A manuscript with careful drawings of this pump and other kinds of instruments from the collection, as well as descriptions of lectures, is still preserved.[28]

At Uppsala experimental physics had been promoted by Andreas Drossander, who acquired many instruments for his demonstrations at the end of the seventeenth century. In 1738, on the initiative of the professor of geometry Samuel Klingenstierna, the University ordered a whole set from London. As yet there was no permanent accommodation but he used rooms at the University hospital. Later, in 1761, the professor of physics Samuel Duraeus bought instruments from Pieter van Musschenbroek's private estate at an auction in Leiden to extend the collection.

---

[27] Jan Larsson, "Den Triewaldska samlingen vid fysiska institutionen i Lund", *Daedalus* 3 (1984), pp. 133–140; John Gudbrand Tandberg, *Die Triewaldsche Sammlung am Physikal. Institut der Universität zu Lund und die Original-Luftpumpe Guericke*, Lund and Leipzig, 1920; J.G. Tandberg, "Historiska instrument i Lund", *Kosmos* 2 (1922), pp. 194–211.

[28] Daniel Menlös, Niclas Schenmark, *Collegium curiosum & experimentale…*, 1743: Uppsala, University Library, UUB A 200.

He also had the collection moved to his private house; in 1788 it was moved to the astronomical observatory; and in 1790, under the professorship of Zacharias Nordmark, into the main university building.[29] The remaining collection is preserved at Museum Gustavianum in Uppsala, but some instruments may still be located in other departments of the University.[30]

Scientific instruments were also collected at the observatories in Stockholm, Lund and Uppsala, along with smaller ones such as at Skara. Even if Stockholm Observatory was originally planned to accommodate space for physical experiments, it should be pointed out that there is as yet no evidence that a cabinet of experimental physics was kept there. However, until the employment of Wilcke in 1759, it was the duty of the astronomer at Stockholm Observatory to teach experimental physics at Riddarhuset, on account of his salary being financed by the Thamian bequest.

In Stockholm the general public was allowed access to the *Laboratorium Mechanicum*, later renamed the "Royal Model Cabinet", initially formed by Christoph Polhem. It was properly a collection of models, but according to the inventory it also contained some scientific instruments. This collection was first displayed at Kungsholmen, moved to different locations, and in 1757 to 1802 displayed at Wrangelska Palatset, not far from Riddarhuset. Triewald borrowed instruments and models from this collection for his teaching. Today the collection of models is preserved at Tekniska Museet in Stockholm. In the capital, physics demonstrations could also take place at the *Collegium Medicum* and the Board of Mines.

When Wilcke argued for the acquisition of new instruments in 1761, he was articulate about defining the space allotted to a proper collection of physics instruments. He wrote that "Astronomy lives in its costly observatories, chemistry in its well-established laboratories, botany and natural history in royal palaces and their gardens, so great that there is hardly their like, mechanics has its model-chambers, medicine is nurtured with the most tender care".[31] Only physics needed to be revitalised: instru-

---

29 Arne Eld Sandström, "The Uppsala Cabinet of Physics", *Annales Academiae Regiae Scientiarum Upsaliensis* 25 (1983–84), pp. 5–46; Staffan Andersson, "Acquisition of a Collection in Context: A Case Study of the Uppsala Cabinet of Physics", in Ewa Wyka, Maciej Kluza, Anna Karolina Zawada (eds.), *East and West: The Common European Heritage. Proceedings of the XXV Scientific Instrument Symposium*, Kraków, 2006, pp. 203–208.

30 At present there is a project to care for and make inventories of the collections of all different departments of Uppsala University in order to get a grip of the universities total collections. Eventually it will be possible to search for artefacts from all departments.

31 Wilcke, op. cit. (n. 3), p. 37.

ments should not only be bought and displayed in a suitable cabinet, but the space should also allow handling and experiments. In the same lecture, Wilcke lamented the lack of state or communal cabinets of physics for public use and teaching, that access to good quality instruments was too often dependent on private collectors, and that these collections were often dispersed after their owners' deaths. He took the opportunity to argue for the establishment of a new and more appropriate cabinet of physics, funded by a communal body in the capital, Stockholm.[32]

There were royal collections, but it is not known to which degree access was granted. At the end of the seventeenth century, some instruments were displayed in the Royal Library at Stockholm Castle and only a few could be saved from the devastating great fire in 1697.[33] The royal collections still hold a few spectacular items from the sack of Prague, but scientific instruments from the royal collections are now also preserved in two different museums: Livrustkammaren (the Royal Armory) and Nordiska Museet. King Adolf Fredrik, who was very interested in the natural sciences and also in handicrafts, had a mathematical cabinet around the middle of the eighteenth century. After his death in 1771 the instruments were auctioned and twenty – many of them made by Daniel Ekström – were presented to the Royal Swedish Academy of Sciences, where they are still preserved.[34]

Besides the royal collections, few notable private collections of scientific instruments existed before the eighteenth century, the best-known being the collection at Skokloster Castle. Knowledge about private eighteenth-century scientific collecting is however scant.

Conclusion

During the eighteenth century interest in experimental natural philosophy boomed in Sweden. This was due not least to the popular teaching performed by Triewald and Wilcke at Riddarhuset in Stockholm, but also to lectures conducted in a more academic context at the Universities in Uppsala and Lund. The lectures were part of a European trend and, while some Swedish lecturers gained knowledge and experience through

32 Ibid.

33 Olof Granberg, "Huru det såg ut i Kongl. Biblioteket två år före Stockholms slotts brand", *Nordisk tidskrift för bok- och biblioteksväsen* 7 (1921), pp. 123–130.

34 RSAS-CfVH: Minutes, 5 February 1772.

travelling in Northern Europe and England, the many publications in the field were valuable also for others who were less mobile. Wilcke claimed that the demonstration of experiments and instruments would draw larger audiences to natural philosophy. He admitted that this would not always provide the auditors with profound knowledge, but it could attract the young and stimulate curiosity within those who had never thought about the subject. Just to awaken ten such persons to natural philosophy would to him be worth more than the cost of the most precious collection of instruments.[35] With such aspirations the teaching of natural experimental philosophy should be both pleasant and useful.

---

[35] Wilcke, op. cit. (n. 3), pp. 33 ff.

DESIGNING THE HOUSE OF KNOWLEDGE IN EIGHTEENTH-CENTURY NAPLES: THE EPHEMERAL MUSEUM OF FERDINANDO SPINELLI, PRINCE OF TARSIA

Paola Bertucci*

The library and museum of Ferdinando Vincenzo Spinelli (1691–1753), Prince of Tarsia, opened to the public in 1747 and soon became one of the most attractive Neapolitan destinations for educated travellers and "Grand Tourists". It was a space for learning and display that defied all classification: it was more than a library, but it could not be defined as a physics cabinet of the likes of Padua, Turin, or Bologna. It was the property of one man, but it was open to the public. It was Neapolitan, but it enjoyed an international reputation. Although not a *Wunderkammer*, the display principles that governed the spatial disposition of the prince's collections were strikingly similar to those that shaped earlier cabinets of curiosities: manuscripts, books, artworks and instruments were gathered together in its rooms as the princely collections of the sixteenth and seventeenth centuries. Yet, the space served as a house for the advancement of knowledge, in the manner of contemporary scientific academies in Paris, London, Bologna and other European capitals. The singularity of this space was evident to contemporaries, who referred to it variously as a library (its official name was "*Biblioteca Spinella*") or as a museum ("Tarsia Museum"), and regarded it as a wonderful temple of knowledge or an outdated ostentation of luxury.

This essay investigates the symbolic significance and the cultural functions of the scientific instruments exhibited in this place of polite sociability, representation and experimentation. Its starting point is the consideration that displaying collections entails a manipulation of space that is shaped by contemporary conceptions of knowledge and of the social

* This essay is a shorter version of Paola Bertucci, "The architecture of knowledge: science, collecting, and display at the Museo Tarsia", in Melissa Calaresu, Helen Hill (eds.), *New Approaches to Neapolitan Culture c. 1500–1800: The Power of Place*, Aldershot, 2013, forthcoming. I am grateful to Helen Hills and Melissa Calaresu for granting me permission to publish this version and for their comments on previous drafts.

functions it should serve. However, the analysis that I offer here moves beyond the notion of the museum as a space for the representation of knowledge. I show that the Tarsia Museum was also a space for cultural experimentation and self-promotion. By analysing its vicissitudes, I wish to point to the tensions arising from contrasting visions of the museum as a space of representation and as a space of knowledge production.

Architecture of Display

Palazzo Tarsia was designed by the most celebrated Neapolitan architect, Domenico Antonio Vaccaro, upon the commission of Ferdinando Vincenzo Spinelli, Prince of Tarsia.[1] As Knights of Malta since the seventeenth century, and marquises of the Holy Roman Empire, the Spinelli were one of the wealthiest and most powerful families in Naples. The building of the *Palazzo* absorbed Spinelli for decades. He began to invest in the project in the 1730s and died before it was finished. The prince worked closely with Vaccaro in the process of designing the building and in the supervision of a constellation of artists, artisans and skilled labourers selected from among the best Naples had to offer. Ambitious as it was, *Palazzo* Tarsia was never completed. The data offered by architectural evidence and archival records are fragmented, yet an engraving by Antonio Baldi on Vaccaro's design offers an extraordinary – if virtual – vision of the project in all its magnificence (*Fig. 1*). By a clever alteration of the perspective, the engraving exalts the sense of splendour and architectural grandeur that Spinelli wanted his palace to materialize. It presents a bird's eye view of the *Palazzo*, complete with annexes, gardens and decorations, most of which were still to be built at the time of publication (1737). Vaccaro's print was an exquisite manifesto of the prince's political and cultural ambitions. It is no accident that Spinelli commissioned it from Vaccaro in 1735, the year of the crowning of the Bourbon King Charles. Until then, Naples had been a province under the viceroys, with no king

[1] Several recent works have discussed *Palazzo* Tarsia from the perspective of the history of art, history of architecture and the history of libraries, but they have only quickly mentioned Spinelli's collection of scientific instruments: Vincenzo Rizzo, *Ferdinando Spinelli di Tarsia. Un principe napoletano di respiro europeo (1685–1753)*, Aversa, 1997; Elena Manzo, *La merveille dei principi Spinelli di Tarsia: Architettura e artificio a Pontecorvo*, Naples, 1997; Vincenzo Trombetta, *Storia e cultura delle biblioteche napoletane: Librerie private, istituzioni francesi e borboniche, strutture postunitarie*, Naples, 2002.

Fig. 1. Domenico Antonio Vaccaro, *Prospetto del Gran Palazzo di Sua Eccellenza il Principe di Tarsia* (Courtesy of the Biblioteca Napoletana di Storia Patria).

living in the city. Charles's decision to settle in Naples aroused great hopes of cultural rebirth amongst the local nobility and marked a significant change in the city's cultural life: in the span of a few years the king promoted the reorganization of the university system and the digging of the archaeological site of Herculaneum, founded new academies and opened the San Carlo theatre.[2] The sparkling renovation of *Palazzo* Tarsia constituted an astounding example of the enthusiasm that the arrival of the king excited in the Neapolitan aristocracy. It was a material expression of support for the Bourbon rule on the part of a family that had a centuries-old history of royal connections. In 1738 Spinelli became a member of the Reale Ordine di San Gennaro, a knightly order instituted the same year by King Charles. The knighthood was granted to a selected elite of local aristocrats particularly faithful to the monarchy, who obtained various religious and civil privileges. The fact that Spinelli was selected to be among the first members of the order testifies to the success of his efforts to signal his support of the Bourbon rule.

2 Raffaele Ajello, "La vita politica napoletana sotto Carlo di Borbone", in Luigi Labruna (ed.), *Storia di Napoli*, Naples, 1971–1978, 10 vols.; v. 7, pp. 961–984.

Palazzo Tarsia was a celebration of the prince's position in Neapolitan society and of his support of the new king. The architecture of aristocratic residences materialized social hierarchy and notions of power both by displaying the symbols of eminence and rank and by allowing selective access to the various areas of the estate.[3] Spinelli and Vaccaro designed the *Palazzo* as a marvel of nature and art, where luxury, imagination, knowledge and technical skills were conjured together to overwhelm visitors with a sense of victory over limits and constraints. Its location in the Pontecorvo, an area recently taken up by local aristocrats, offered breathtaking views of the Gulf of Naples and Mount Vesuvius. While admiring the valuable collection of paintings in sumptuous reception rooms, the prince's most illustrious guests would simultaneously enjoy the most emblematic vistas of the city.[4]

The prince devoted a large section of his palace to learning and display in a series of rooms that than came to be called the "*Biblioteca Spinella*" or "Tarsia Museum", in which he placed books, artworks and scientific instruments.[5] It was the fame and reputation of this museum and library, open to the public three times a week, that made *Palazzo* Tarsia renowned in the republic of letters and a destination for Grand tourists and learned travellers arriving in Naples.

The Temple of Minerva in Naples

The *Biblioteca Spinella* corresponded in significant ways to the ideal library described by Gabriel Naudé, Cardinal Mazarin's librarian, in his *Advice on establishing a library*.[6] As the "*Biblioteca Spinella*", the library of *Palazzo* Tarsia blended its name and reputation with that of its owner. It had an

---

[3] Norbert Elias, *The Court Society*, New York, 1983; Chandra Mukerji, *Territorial ambitions and the gardens of Versailles*, Cambridge, 1997; Gérard Labrot, *Palazzi napoletani: Storie di nobili e cortigiani*, Naples, 1993.

[4] Spinelli owned an impressive collection of paintings, which included works by Giotto, Raphael, Titian, Michelangelo, Andrea del Sarto, Veronese, Caravaggio, Carracci, Rubens, Van Dyck, Guido Reni, Tintoretto among many others. For a full list of painters, see Carlo Celano, *Delle notize del bello, dell'antico, e del curioso della città di Napoli per li signori forastieri*, Naples, 1792, 4th ed., 10 vols.; v. 6, pp. 61–62.

[5] According to Vaccaro's captions in the engraving of Palazzo Tarsia, the library was 138 × 46 Neapolitan *palmi* (approximately 36 x 12 meters). This contrasts with the larger measurements recorded in later descriptions of the actual library (Vaccaro's was a project), which I will use in the remaining of the paper. This might indicate that the library – and the collection of instruments – acquired more importance as the building progressed.

[6] Gabriel Naudé, *Advis pour dresser une bibliothèque*, Paris, 1644.

independent entrance, which distinguished readers and visitors from the prince's guests, and it was composed of four rooms detached from the palace's main body, with two galleries that overlooked the beautiful gardens. The prince spared no financial outlay in making his library pleasant to the eye. He spent considerably on *indorature*, the gilding that coated all the banisters, frames, stuccoes and other decorative elements in the *Palazzo*. The noblest metal, symbol of opulence and prosperity, was also deployed generously on the leather bindings of books and on the carved cabinets that housed them. Indeed, the abundance of gilding in the library was dazzling and, in contemporary descriptions, became one of its most striking features.

The list of expenses that the prince sustained reveals his desire to spare nothing to produce a magnificent house of knowledge, for the cultivation of science, culture and the arts.[7] It was to be a library as well as a "museum", in the literal sense of a temple of the muses, where the learned would find inspiration. The heterogeneity of the items displayed in the *Biblioteca Spinella* reminded contemporaries of the collections of the Medici in Florence, where objects of art and objects of science coexisted.[8] However, the Tarsia Museum was not simply the product of a collecting activity guided by learned curiosity, nor was it simply a display of connoisseurship styled upon earlier princely cabinets or *Wunderkammern*. Instead, Ferdinando Spinelli entrusted the gathering of books and of scientific instruments to local dealers and scholars, some of whom furnished him with pre-existing book collections.[9] He purchased a valuable collection of seventeenth-century mathematical instruments from the bookseller Porcelli and commissioned from a Dutch instrument-maker, most probably Pieter van Musschenbroek, the manufacture of several physics instruments for experimental demonstrations.[10] More than simply representing Spinelli's involvement in culture, his library and museum were a statement about his intention to create in Naples a space where cultural activities could be carried out under his (and the king's) patronage.

Ideal architectural structures were often invoked in early-modern writings on human learning. In the seventeenth century, the biblical images of the garden, the ark, the tower and the temple were commonly referred

[7] Rizzo, op. cit. (n. 1).
[8] Gherardo Degli Angioli (De Angelis), *Orazioni. Parte I*, Naples, 4th ed. 1763, pp. 276–277.
[9] Trombetta, op. cit. (n. 1), pp. 100–102.
[10] Rizzo, op. cit. (n. 1), p. 109.

to as metaphors of knowledge.[11] Spinelli resorted to classical imagery: in the wake of the enthusiasm for the ancient roots of Naples, triggered by the digging of Herculaneum, he designed his museum and library as the Temple of Minerva – the goddess of knowledge born from Jupiter's brain.[12] An epigram composed by the philosopher Giambattista Vico, engraved in gold above the entrance door, greeted visitors to the Temple, sealing the relationship between knowledge and power that the *Biblioteca Spinella* materialized: "Born in heaven from Jupiter's brain, Minerva / Dwells here on earth in golden shelters worthy of Jupiter".[13]

Minerva was the learned offspring of the most powerful of the gods. Her sacred abode on earth was a site that invited the respectful celebration of ancient cultures and called for the advancement of knowledge. Similar to Francis Bacon's House of Solomon, the Temple of Minerva was a site for observation, learning and experiment, yet it did not exist on Bacon's imaginary Bensalem, but in the flourishing Bourbon city of Naples. It was there that learned élites could now find a new patron for the sciences and the arts.

The *Biblioteca Spinella* was simultaneously a Neapolitan object and a cosmopolitan place. It represented the prince's vision of a centre of learned sociability that connected Naples to other European cultural capitals, under the aegis of the Spinelli family and of the Bourbon dynasty. Whether travellers or scholars, visitors who entered the Tarsia Museum and Library walked through a series of rooms that represented to them the position of Ferdinando Spinelli and his family in the cultural history of Naples and the central role that the new site of knowledge would play in the cultural awakening to follow the arrival of the Bourbon king.

The library's first, and largest, room housed Latin, Greek and Italian books mostly bound in Neapolitan style.[14] The portrait of Ferdinando Spinelli was hung in a central position above its main door, reminding readers who entered through the independent entrance that the Temple of Minerva was part of *Palazzo* Tarsia. At the opposite end of the room, facing Ferdinando's portrait, hung that of his father, whose book

---

[11] Jim A. Bennett, Scott Mandelbrote, *The Garden, the Ark, the Tower, the Temple: Biblical Metaphors of Knowledge in Early Modern Europe*, Oxford, 1998.

[12] On antiquarianism in eighteenth-century Naples see the special issue "Antiquarianism, museum, and cultural heritage. Collecting and its contexts in eighteenth-century Naples", *Journal of the history of collections* 19 (2007).

[13] Placido Troyli, *Istoria Generale del Reame di Napoli*, Naples, 1752, p. 141. My translation.

[14] Ibid. According to Celano, this room was 120 × 40 Neapolitan *palmi* (approximately 31.1 × 10.5 meters): Celano, op. cit. (n. 4), v. 6, pp. 70–72.

collection constituted the original core of the library. The prince used portraits of himself and his family to situate the Temple of Minerva within the Spinelli family's cultural patronage in Naples. Marble statues of each of the four seasons standing in the room's four corners, reminded visitors of the course of time, which culminated with Naples' alliance with the Spanish crown. Such alliance was materially represented by a remarkable pyramidal cabinet designed especially by local artisans, placed at the centre of the room. Made of ebony, turtle shell and gold-plated copper, this quintessential symbol of hierarchy and power stood upon a carved gilt pedestal and was crowned by a double clock that gave Italian and Spanish time.[15] Two armed loadstones, symbols of the force of attraction that draws people together, were placed on the sides of the pyramid, which served as a display case for an exquisite collection of seventeenth-century mathematical instruments made by the French instrument-maker Nicholas Blondeau.[16] The collection (that included gold and silver items) had belonged to a previous Neapolitan collector who had sold it to Porcelli, for re-sale. It was unusual for eighteenth-century collectors to display historical instruments, and contemporary descriptions of the Tarsia Museum did not fail to emphasise the fine craftsmanship of "the distinguished and famous Blondon".[17] In placing it at the centre of the main room in an attractive display case, Spinelli set the scientific activities of the museum in continuity with previous local traditions. Along the corridor, visitors could admire a terrestrial and a celestial globe made by Vincenzo Coronelli. Other expensive astronomical instruments, such as a valuable astronomical quadrant, signed by the English maker Sisson, and a semicircle with telescopic sights, were also on display in this room. This was probably on a temporary basis only, as according to Vaccaro's drawing a "noble and decorated observatory" was to crown the entire palace

[15] The clock was made by two English clockmakers who worked for the prince, Oliver Robinson and Charles Tredman. See Rizzo, op. cit. (n. 1), p. 10.

[16] Blondeau was a French maker who moved to Milan in the late seventeenth century and operated in Naples around 1694. I am grateful to Anthony Turner who helped me identify in Nicholas Blondeau the otherwise unknown Nicola Blondone or Blondel mentioned in eighteenth-century descriptions of the *Biblioteca Spinella*. It is likely that an existing sector signed "Blondeau, Naples" made in 1694 was among the instruments purchased by the prince of Tarsia from Porcelli.

[17] Troyli, op. cit. (n. 13), p. 241. On eighteenth-century collectors' preference for new instruments, see Anthony J. Turner, "From mathematical practice to the history of science", *Journal of the history of collections* 7 (1995), pp. 135–150.

(*Fig. 1, letter Q*).[18] When the French astronomer Jerôme de Lalande visited *Palazzo* Tarsia in 1766, however, he described its astronomical instruments, but did not mention an observatory.[19] It is likely that the death of the prince in 1753 prevented the completion of this project. Indeed, no description of the observatory exists.

The library's second room was dedicated to foreign cultures. It held foreign books with French bindings, and its central piece was an English orrery that could be modified to illustrate the geocentric system.[20] A series of portraits of the Spinelli princes, arranged according to their genealogy starting with their tenth-century progenitor, decorated the walls around the orrery. Above the door leading to the next room an epigraph engraved in gold letters on "touchstone" glorified Ferdinando Spinelli for making his collections available to the public.

Access to the working areas of the library implied an obliged choreography through these two rooms, in which knowledge was represented as emanating from Spinelli's munificence. Readers had no other way to reach the reading room, where they could sit on chairs covered with crimson silk embroidered with gold, and work on Chinese desks coated with black leather and gilded decorations. Here, portraits of illustrious men hung from the walls. Between the reading room and the physics cabinet (the fourth room) there was a gallery where two local savants, the astronomer Felice Sabatelli and the naturalist Giovanni Maria Della Torre, built a meridian line modelled on those made by Gian Domenico Cassini for the Church of San Petronio in Bologna and by Francesco Bianchini for the Church of Santa Maria degli Angeli in Rome. The works for the meridian, according to an epigraph engraved in it, started in 1749 and were completed in 1759.[21]

---

[18] It should be noted that the observatory in Vaccaro's design seems more an idealization than a realistic project. See Troyli, op. cit. (n. 13), p. 241, and Tommaso Di Costanzo, *Ferdinandi Vincentii Spinellii Tarsiae principis bibliothecae index alphabeticus secundum authorum cognomina dispositus*, Naples, 1780, p. 128. Troyli erroneously describes the quadrant as a Dutch instrument, and lists also an electrical machine in this room, which appears in the second room in Di Costanzo's inventory. The differences maybe due to the different dates of publication.

[19] Joseph-Jérôme de Lalande, *Voyage d' un François en Italie, Fait dans les années 1765 et 1766*, Paris, 1769, 8 vols.; v. 6, pp. 199–200. The observatory is not mentioned in later revised editions of this work either.

[20] Troyli, op. cit. (n. 13), p. 141. According to Celano this room was 50 × 20 *palmi* (approximately 13 × 5.2 meters): Celano, op. cit. (n. 4), v. 6, pp. 61–68.

[21] Troyli, op. cit. (n. 13), p. 141. On meridians, see John Heilbron, *The Sun in the Church: Cathedrals as Observatories*, Cambridge (Massachusetts), 1999.

The gallery marked the transition from spaces of study (reading room) and display (first two rooms) to one of exhibition and performance: the physics cabinet. In this room, which hosted the collection of philosophical instruments, the prince and his guests enjoyed the spectacle of experiments. Spinelli owned the most representative machines of experimental philosophy, covering all branches of mid-eighteenth-century experimental philosophy: mechanics, astronomy, optics, electricity, magnetism, hydrostatics and pneumatics.[22] In line with the international character of contemporary science, the instruments came from a variety of places. Most of them were made in the famous workshop of Jan van Musschenbroek, the Dutch instrument maker who furnished the physics cabinets of the University of Padua and of the Istituto delle Science in Bologna. A few items carried the signature of renowned English makers and others were manufactured locally.[23] The cosmopolitan nature of this space was emphasized by four marble statues of the known continents placed in the four corners, similar to the arrangement of the library's first room. Architectural drawings and engraved cameos embellished the walls with *vedute* of the useful applications of the new philosophy, while silk *chaises longues* were available to the prince's guests for the comfortable enjoyment of the spectacle of science. The decorative pattern of this room accorded with the rest of the *Biblioteca Spinella*: crimson silk covered walls and chairs,

22 From the records of Spinelli's expenses published in Rizzo, op. cit. (n. 1), we can provide a chronology of a number of instruments purchased by the prince of Tarsia: the "machines from Holland" arrived in 1745, when he requested the master woodcarver Ponsiglione to make decorated armoires to house them; in 1746 he acquired from the bookseller Porcelli Blondeau's mathematical instruments, a brass armillary sphere for the Copernican system and a Newtonian telescope; in 1747 he purchased two armed loadstones and commissioned a balcony for a camera obscura and a "newly invented" microscope (probably a solar microscope) from the carpenter Marazzo; in 1748 the gilder Pane varnished and decorated an electrical machine. Musschenbroek often acted as a retailer for instruments made in England, so it is likely that the orrery, the Sisson quadrant and the other astronomical instruments arrived from Holland in 1745. In 1780 the lawyer Di Costanzo published an inventory of the books from the library of the Prince of Tarsia. At the end of the volume there is a list of Spinelli's scientific instruments; comparing it with the description of the library by Troyli (1752) it appears that no significant additions were made, see Di Costanzo, op. cit. (n. 18); Troyli, op. cit. (n. 13), pp. 241–244.

23 On the Musschenbroek workshop, see Peter R. De Clercq, *At the Sign of the Oriental Lamp: The Musschenbroek Workshop in Leiden, 1660–1750*, Rotterdam, 1997. De Clercq mentions a commission to Musschenbroek in 1744 for a "distinguished place in Italy" (ibid., p. 169); it is possible that it was *Palazzo* Tarsia.

the wooden cabinets were gilt and carved as well as the frames for the paintings on the walls.[24]

Unlike other rooms in the library, the instruments on display in the physics cabinet could be taken off the shelves and used to stage spectacular experimental demonstrations. In larger physics cabinets, such as those of the University of Turin or of the Bologna Institute of Sciences, instruments were assembled together according to the particular branch of experimental philosophy they illustrated: mechanics, hydraulics, pneumatics, optics, electricity or magnetism. In the Tarsia Museum, instead, the philosophical instruments were arranged according to size, as were books in most seventeenth-century libraries: smaller items were shelved in wooden cabinets, while the largest ones were placed in the centre of the room.[25] The coexistence of new cultural trends and older modes of display was a distinctive feature of the Tarsia Museum; it reflected also in how the prince saw himself in relation to the activities that took place in his physics cabinet.

Old Models for a New Space of Knowledge Production

Itinerant lecturers who made their living by the visual demonstration of the laws of nature contributed largely to the success of eighteenth-century experimental philosophy across Europe. The vast array of "philosophical" instruments employed in their spectacular lectures rapidly became cultural commodities and new objects of collecting. The most expensive collections of instruments were found in institutions such as universities and academies, but there were also individual savants who spent fortunes on their physics cabinets. Thanks to the emerging taste for "philosophical collections", instrument makers in Leiden, Paris and London became international celebrities, with customers in Europe and its colonies, while lesser known makers obtained status and prestige at the local level. As with cabinets of curiosities and *Wunderkammern*, the instrument trade created a certain degree of uniformity in these collections: items such as the air pump and the electrical machine were almost fetishised, and even

[24] Troyli, op. cit. (n. 13), p. 243. According to Celano, this room was 40 × 18 *palmi* (approximately 10.4 × 4.7 meters); Celano, op. cit. (n. 4), v. 6, pp. 61–68.

[25] On book arrangement in early modern libraries, see Eric Garberson, "Libraries, Memory and the Space of Knowledge", *Journal of the History of Collections* 18 (2006), pp. 105–136.

the most parsimonious of collectors would have a few brass instruments signed by a famous English maker. Thanks also to the successful textbooks that instruments makers published, the names of George Adams, Edward Nairne, Jesse Ramsden, Jeremiah Sisson, became synonymous with a widespread aesthetic of precision that turned scientific instruments into luxury items for the learned.

Physics cabinets exhibited more than skilful craftsmanship, however. Philosophical instruments on display represented a new manipulative approach to natural knowledge that distinguished the cabinet of experimental philosophy from previous forms of scientific collecting. They were the means by which the spectacle of nature could be recreated at will, the totemic embodiments of natural powers tamed and put to use for the "benefit of mankind". Philosophical instruments on display could be admired *per se*, as icons of the new natural knowledge founded on experiment, but more interestingly they could be taken off the shelves and used for unprecedented forms of spectacle that blended the domains of the natural and of the artificial.[26]

The theatrical nature of experimental philosophy elicited a variety of responses on the part of audiences, ranging from enjoyment of a new form of spectacle to participation in experimental research. The physics cabinet of Ferdinando Spinelli played a double role as a stage for amusing performances as well as a site where a group of Neapolitan savants transformed themselves from spectators of experimental demonstrations to producers of new knowledge. This crucial transformation stood at the basis of the creation of a scientific academy that in a short span of time made visible the contrasting cultural visions that met in the *Biblioteca Spinella*: that of the prince of Tarsia and that of the academicians.

Spinelli was not directly involved in the new natural philosophy; he sponsored the experimental activities that took place in his physics cabinet by fashioning himself as a patron of the sciences. In doing so, he emulated the powerful Florentine family of the Medici, who owned a fine collection of mathematical instruments, on display in the "*Stanzino delle matematiche*" (Room of mathematics) of the Uffizi Gallery, and supported the activities of the first experimental academy, the "*Accademia del Cimento*",

---

26 See Larry Stewart, *The Rise of Public Science. Rhetoric, Technology and Natural Philosophy in Newtonian Britain, 1660–1750*, Cambridge, 1992; Jan Golinski, *Science as Public Culture. Chemistry and Enlightenment in Britain, 1760–1820*, Cambridge, 1992; Simon Schaffer, "Natural Philosophy and Public Spectacle in the Eighteenth Century", *History of Science* 21 (1983), pp. 1–43.

which gathered in the prince's palace. Spinelli invested in his scientific collection during a short span of time, around 1745, most likely as a result of the closure, one year earlier, of the Neapolitan Academy of Science, which had been founded by Celestino Galiani in 1732.[27] His purchase of seventeenth-century mathematical instruments and their ostentatious display in the *Biblioteca Spinella* was an indication of his wish to support scientific activities in continuity with earlier attempts.[28] Indeed, in 1747 Spinelli encouraged the formation of a new academy, tellingly named "*Accademia Spinella*", which gathered in his museum. Its organization was due to the Somaschan cleric Giovanni Maria Della Torre, a physics professor at a local religious school, who also worked as the curator of the physics cabinet. Della Torre, whose duties were similar to those of the librarian, was responsible for the acquisition of new instruments and for the staging of experimental demonstrations to entertain the prince's guests.[29] He gave a clear Newtonian orientation to the physics cabinet of *Palazzo* Tarsia and enriched the collection with a Newtonian telescope and several prisms, in addition to the large orrery on display in the library's second room.

The Physics Cabinet as a Space of Self-Promotion

Focusing on the acquisition practices that led to the formation of specific collections, historians of scientific instruments have cast light on the extensive circulation of instruments related to the emergence of the physics cabinet in the eighteenth century.[30] I would like to suggest that this new space of display, which was often a place where experimental

[27] Rizzo, op. cit. (n. 1), *passim*. Vaccaro's engraving does not mention any instrument in the description of the library. Although in 1749 Spinelli discussed with the *abbé* Nollet the possibility of enlarging his physics cabinet, there is no evidence that he actually commissioned any item from the French maker (see n. 29 below).

[28] Eighteenth-century collections of scientific instruments were self-consciously "modern" and, normally, did not include older objects, even if they had belonged to celebrities such as Isaac Newton; see Turner, "From mathematical practice…", cit. (n. 17), pp. 135–150.

[29] Jean Antoine Nollet, "Journal du voyage de Piémont et d'Italie en 1749": Soisson, Bibliothèque Municipale, Ms 150, f. 176.

[30] See, for example, Alan Q. Morton, Jane Wess, *Public and Private Science: The King George III Collection*, Oxford, 1993; Lewis Pyenson, Jean-François Gauvin (eds.), *The Art of Teaching Physics. The Eighteenth-Century Demonstration Apparatus of Jean Antoine Nollet*, Sillery, Quebec, 2002; P. Bertucci, "Public utility and spectacular display: The physics cabinet of the Royal Museum in Florence", *Nuncius* 21 (2006), pp. 323–336.

activities took place, produced also new forms of social interaction. The physics cabinet of *Palazzo* Tarsia proved crucial for the self-promotion of the people who gathered there. Thanks to its collection of instruments and to the ambitions of Della Torre, the physics cabinet of *Palazzo* Tarsia attracted several savants interested in pursuing experimental research, including Mariangela Ardinghelli, the Italian translator of Stephen Hales' Newtonian textbooks.[31] Their activities quickly focused on electricity, a new experimental science that in the course of 1746 became extremely fashionable, in the Italian countries as in the rest of Europe.[32] After the thrilling performances in *Palazzo* Tarsia of Johannes Petrus Windler, a Saxon electrical showman, Spinelli commissioned a big electrical machine from a local maker, had it varnished and gilded in accordance with the library's style, and entrusted it to Della Torre.[33] For the prince, the electrical machine became a marvel with which to entertain his guests, whereas for the academicians it became an opportunity to boost their visibility in the Neapolitan cultural scene (*Fig. 2*).[34]

For a short time, Spinelli's ambition to act as a patron of the sciences in the style of Leopoldo de' Medici went hand in hand with the expectations of the savants who gathered in the Tarsia Museum. They found in him a protector who sponsored their activities and, in exchange, their work made visible his commitment to the advancement of knowledge. This momentary convergence of interests is epitomized in *Tentamina de causa electricitatis*, a book that described Windler's experiments in the Tarsia Museum visually as well as textually, published by Della Torre in 1747. The book was a public celebration of Spinelli's patronage of the sciences, with a long dedication exalting the prince of Tarsia and his library as the perfect place for experimental research.[35]

---

[31] On Ardinghelli, see P. Bertucci, "The in/visible woman: Mariangela Ardinghelli and the circulation of knowledge between Paris and Naples in the eighteenth century", *Isis* (June 2013), forthcoming.

[32] I have detailed the peregrination of the electrical machine in the Italian peninsula in P. Bertucci, *Viaggio nel paese delle meraviglie. Scienza e curiosità nell'Italia del Settecento*, Turin, 2007, pp. 101–122 and 151–240.

[33] Rizzo, op. cit. (n. 1), p. 121.

[34] Giovanni Maria Della Torre, *Scienza della Natura*, Naples, 1748, 2 vols.; v. 2, p. 316. In the longer version of this paper, Bertucci, "The architecture of knowledge…", cit. (n. *), forthcoming, I explore the activities of the savants who joined the academy in the Tarsia Museum. One of them was the Neapolitan *femme savante* Mariangela Ardinghelli; see Bertucci, "The in/visible woman"…, cit. (n. 31), forthcoming.

[35] Petrus Johannes Windler, *Tentamina de causa electricitatis*, Naples, 1747, unpaginated dedication. For an analysis of the significance of this work in the context of the

Fig. 2. Electrical experiments in Palazzo Tarsia; Joannes Petrus Windler, *Tentamina de causa electricitatis*, Naples, 1747 (Courtesy of the Bakken Museum and Library of Electricity in Life).

The Ephemeral Academy

Although experimental activities were well under way in the Tarsia Museum, Spinelli waited until 22 July 1747 to open the library formally to the public. The memorable ceremony, which cost him as much as his

---

Accademia Spinella and of Della Torre's role in the publication see the longer version of this paper. [forthcoming]

daughter's wedding, was delayed until the birth of Filippo, the crown prince.[36] A few weeks later, Spinelli pursued his emulation of the Medici by constituting the *Accademia Spinella*, an assembly of local savants who gathered in his palace under his patronage. The members of the academy embraced Spinelli's support of the Bourbon monarchy. One of their first formal acts was the notification of the newly-formed institution to the Académie Royale des Sciences in Paris. In an official communication to Georges Louis-Leclerc, Comte de Buffon, they expressed their desire to engage in "scientific commerce" with their French colleagues. They hoped that the former French ambassador in Naples, the Marquis de l'Hôpital, would be willing to act as a mediator. They asserted that their primary concern was with natural history, knowing that the study of Mount Vesuvius and of places such as the Campi Phlegraei was of great interest to Buffon and, more generally, to the international republic of letters. Clearly, the attempt to attract the interest of the Paris Academy had to do not only with the prestige of the institution, but also with Spinelli's links with the monarchy: it was a branch of the Bourbon family that ruled France.

However, the Neapolitans' letter to Buffon revealed ambitions that contrasted with the old model of the "prince's academy" dear to Spinelli. They manifested their intention to shape the *Accademia Spinella* after its fellow institution in Paris, a striking statement when one considers that the French members received a salary from the king.[37] The attempt to create a scientific academy in Naples modelled on the Académie Royale des Sciences in Paris had characterized also Galiani's failed project, which the *Accademia Spinella* was aiming to revive. Natural history was the primary interest of Galiani's academy and the document sent to Buffon made explicit the connection with his project: "our still nascent academy [...] after various vicissitudes, and various fortunes, has lately been welcomed by His Excellency Signor Prince of Tarsia".[38] Although no list of members of the new academy exists, Della Torre referred to former members of Galiani's academy as members of the *Accademia Spinella*.[39]

---

36 Rizzo, op. cit. (n. 1); Camillo Minieri Riccio, "Cenno storico delle Accademie fiorite nella città di Napoli", *Archivio Storico per le Province Napoletane* 5 (1880), pp. 349–367: 367.

37 Paris, Archives Académie des Sciences, Procès Verbaux, 1748, p. 31.

38 Ibid. On Galiani's academy, see Vincenzo Ferrone, *Scienza, natura, religione: mondo newtoniano e cultura italiana nel primo Settecento*, Naples, 1982, pp. 457–485.

39 Della Torre, op. cit. (n. 34), v. 2, pp. 159, 250.

Similar though their research agendas might have been, however, the *Accademia Spinella* did not elicit any sympathy on Galiani's part and he never participated in its activities.[40] The Temple of Minerva as envisioned by Spinelli was not attuned to the ideal of knowledge that Galiani and his followers were seeking to establish in Naples. For them, the whimsical patronage of a prince was ill-suited to the advancement of the sciences. The model of the prince's academy was despicable for Galiani and his closest collaborators, who commented sarcastically on the inauguration of the *Biblioteca Spinella*:

> If Signor Prince of Tarsia, with his conceit of *gran signore*, wanted to dispense [genuine] literary magnificence, he could make himself really glorious other than with the opening of the Library with Music and refreshments.[41]

This criticism of the *Biblioteca Spinella* anticipated a crucial divide between Spinelli's and the academicians' understanding of the role of the Temple of Minerva in Naples' cultural life. For the prince it was a place that represented his patronage of the sciences; the academicians hoped that the new knowledge they produced there would secure for them salaried positions in Neapolitan society. However, King Charles did not provide the support that the academicians wished for, condemning the academy to a fleeting existence.[42] In June 1749 Della Torre confessed to a correspondent that the academy was dying and Spinelli himself did not regard it as worth mentioning a few months later, when he had a long conversation about his physics cabinet with the *abbé* Nollet.[43]

Della Torre was not foreign to the failure of the academy. Thanks to the Prince of Tarsia's connections with the royal family and the patronage of Cardinal Spinelli, he obtained various prestigious positions in the newly-founded Reale Accademia Ercolanense, in the Museo Farnesiano, in the Royal Library and in the Royal Printing Press.[44] His successful trajectory, begun in the Tarsia Museum and continued in royal institutions, was emblematic of the contrast between Spinelli's cultural vision and the

---

40 Aldo Brigaglia, Pietro Nastasi, "Bologna e il Regno delle Due Sicilie: Aspetti di un dialogo scientifico (1730–1760)", in Renzo Cremante, Walter Tega (eds), *Scienza e letteratura nella cultura italiana del Settecento*, Bologna, 1984, pp. 211–232.

41 Naples, Biblioteca della Società Napoletana di Storia Patria, XXXI B19, f. 159v.

42 For a more detailed analysis of the reasons leading to the quick end of the academy see the longer version of this paper. [forthcoming]

43 G.M. Della Torre, Letter to Paciaudi, 24 June 1749, cit. in Trombetta, op. cit. (n. 1), p. 92. Nollet, "Journal" . . ., cit. (n. 29), f. 181.

44 See Elvira Chiosi, "Academicians and Academies in Eighteenth Century Naples", *Journal of the History of Collections* 19 (2007), pp. 177–190.

expectations of Neapolitan savants. Della Torre obtained his positions thanks to his participation in Spinelli's project, which allowed him to benefit from the prince's connections with the crown, yet the success of his individual trajectory corresponded to the failure of the collective project of the *Accademia Spinella*. The members' expectations were frustrated by his rapid ascent: after obtaining his appointments, he lost interest in the academy. In spite of professed intentions, his interactions with the Tarsia Museum and with Spinelli unfolded according to a patron-client relationship typical of the court system.

Conclusion

Ferdinando Spinelli envisioned the Temple of Minerva and the *Accademia Spinella* respectively as Neapolitan adaptations of the *Stanzino delle matematiche* of the Uffizi Gallery and of the *Accademia del Cimento*. Yet, while the Medici were the ruling family in Florence, Spinelli had to compete with many other aristocrats for the king's attention in Naples. The *Accademia Spinella* was one of many attractions that *Palazzo* Tarsia offered to the king, who did not deliver the support that the academicians expected. If Spinelli embraced the earlier model of the prince's academy, his academicians held international ambitions, grounded in the ideal of an institutionalized, state-funded place for the promotion of natural philosophy. So, the Temple of Minerva came to be perceived as Spinelli's outdated *Wunderkammer*, a space of ostentation and display subjected to the whimsical inclinations of his powerful, yet uneducated owner who, as a contemporary put it, seemed to appreciate books more for their size than for their contents.[45]

After Spinelli's death in 1753 the library remained open to the public for a couple of decades, until his heirs sold most of its collections to the king.[46] If the *Palazzo* continued to be a destination for Grand tourists, to Neapolitan savants it became the emblem of inconsequential enterprises. The satire entitled *Note of the books from the ephemeral library of the most excellent the gilded prince of Tarsia* indicates the irremediable divergence between the cultural vision of a prince, who imagined himself as a new Maecenas, and that of pragmatically-oriented intellectuals. The *Note* provides a long list of impossible books, whose fake titles, authors,

45 Alfredo Settai, *Domenico Passionei e la sua biblioteca*, Milan, 2004, p. 608.
46 Trombetta, op. cit. (n. 1), pp. 104–106.

and publishers, mock the catalogue of the *Biblioteca Spinella* to sketch a grotesque microcosm of vanity, affectation and pretentiousness.[47] This harsh alternative vision of the *Biblioteca Spinella* testifies to the failure of the architecture of knowledge materialized in Spinelli's Temple of Minerva. Like the fantastic engraving of Vaccaro's design of the *Palazzo* Tarsia, Spinelli's cultural vision came to be perceived more as a frivolous mirage than as a viable project.

[47] "Satira. Nota dei libri Pervenuti dalla Libreria Efimera dell'Ecc.mo Principe di Tarsia Indorato": Naples, Biblioteca della Società Napoletana di Storia Patria, MS XXIII C 5, f. 11. The manuscript is anonymous and undated, yet it seems modeled upon the catalogue of Spinelli's library published in 1780.

BETWEEN TEACHING AND COLLECTING: THE LOST CABINET OF PHYSICS OF PRINCES JOSÉ AND JOÃO OF PORTUGAL (1777–1808)

Marta C. Lourenço and David Felismino*

Introduction

Royal collections have a special place in the history of collections. Kings and princes are known to have assembled collections or proto-collections at least since the Middle Ages.[1] Due to the wealth, social importance and power of their owners, royal collections are among the richest and largest of any given period. Some of the most important museums in Europe were organised around specimens, artefacts and artworks from royal collections. The study of their constitution, development and dispersal offers new perspectives and insights into the history of the sciences and the arts, but also into the development of technological and economical innovation, geopolitics, social relations, taste and fashion, and popular culture.

Among royal collections, cabinets of physics are particularly interesting. Although they share characteristics with other cabinets of physics – say, in universities – they are first and foremost private collections and it is often difficult to establish their precise purpose and scope. They may combine teaching instruments, precision instruments, highly crafted and merely decorative instruments and "applied" instruments materialising state-of-the-art technological and industrial development (often gifts made by traders with economic aims). The multiple layers of royal cabinets of physics remain largely unstudied. Moreover, in Portugal, despite the fact that the court played an important role in gathering scientific instruments since as early as the Renaissance, royal cabinets of physics have received little recognition from Portuguese historians. Literature,

* We are grateful to the Foundation for Science and Technology (Lisbon) for supporting this research and to the sources noted in captions for permission to use their images. We also thank the Secretary of State for Culture for access to the collections and archives of the Portuguese palaces.

[1] Patrick Mauriès, *Cabinets of Curiosities*, London, 2002; Oliver Impey, Arthur MacGregor, *The Origins of Museums. The Cabinet of Curiosities in Sixteenth- and Seventeenth-Century Europe*, London, 2001.

exclusively in Portuguese, is fragmented and based, to a great extent, on limited documental sources.[2]

In this paper, we will present and discuss a particularly illustrative case of the multiplicity of purposes and diversity of scope of royal cabinets of physics – the cabinet organised for Princes José (1761–1788) and João (1767–1826) in the late 1760s at the Palace of Ajuda (Lisbon).[3] Sources used comprise documents, instruments and iconography. In the first section we will provide context about eighteenth-century cabinets of physics in Lisbon, particularly those sponsored by King João V (1689–1750) and, in the second section, we will discuss the aims, composition and use of the princes' cabinet.

The Early 1700s: Cabinets of Physics Sponsored by King João V

It was only in the early 1700s, under King João V, that the abundance of gold from Brazil enabled the development of an intense and prolific "scientific" activity in Lisbon,[4] encompassing cabinets of physics, astronomical observatories and natural-history cabinets (*Fig. 1*). Experimentation was considerably common, reflecting the development of experimental philosophy across Europe.[5] Professionals and amateurs, universities, academies and the royal family collected instruments and used them in cabinets.

[2] The Museum of the University of Lisbon has the largest collection of scientific instruments with royal provenance in Portugal. Since April 2010, we have been attempting to fill in this gap with further research ("*On the Instruments' Trail. Exploring Royal Cabinets of Natural Philosophy in Portugal*", PTDC/HIS-HCT/098970/2008, financed by the FCT-MCTES). For more details, see Marta C. Lourenço, "Royal Cabinets of Physics in Portugal and Brazil: An exploratory study", *Opuscula Musealia* 19 (2012), pp. 71–88; Marta C. Lourenço, David Felismino, "Les cabinets de physique du Roi du Portugal, XVIIIe–XIXe siècle: Organisation, Dispersion et Collections Perdues", in Ana Cardoso de Matos, Marie-Sophie Corcy, Christiane Demeulenaere-Douyère, Irina Gouzévitch (eds.), *Cabinets de curiosités, collections techniques et musées d'arts et métiers*, Paris, forthcoming.

[3] João Pedro Ferro, *Um príncipe iluminado português: D. José (1761–1788)*, Lisbon, 1989, pp. 56–62; Maria Beatriz Nizza Da Silva, *D. João V*, Lisbon, 2006, pp. 122–124; Nuno Gonçalo Monteiro, *D. José. Na sombra de Pombal*, Lisbon, 2006, pp. 250–253.

[4] Charles Boxer, *O Império Marítimo português. 1415–1825*, Lisbon, 1977, pp. 218–219.

[5] Rómulo de Carvalho, *História da Fundação do Colégio Real dos Nobres de Lisboa*, Coimbra, 1959; de Carvalho, *História do Gabinete de Física da Universidade de Coimbra*, Coimbra, 1978; de Carvalho, *A Física Experimental em Portugal no século XVIII*, Lisbon, 1982; de Carvalho, *A astronomia em Portugal no século XVIII*, Lisbon, 1985; de Carvalho, *O material didáctico dos séculos XVIII e XIX do Museu Maynense da Academia das Ciências de Lisboa*, Lisbon, 1993; Maria Emilia V. Gomes, *Desenvolvimento do ensino da física experimental em Portugal 1780–1870*, PhD thesis in Physics, University of Aveiro, 2007.

Fig. 1. King João V (1689–1750), depicted with several scientific instruments in an engraving by Stefano and Rocco Pozzi, Lisbon, 1730s (National Library of Portugal, Lisbon, BNP E. 44V). On the monument: "Ioanni.V. Lusit. Et Algarb. Regi Scienciae Artes Que Auctar Ac Ditatae".

João V was a king of broad sensitivity towards the arts, sciences and humanities. In 1721, he instructed Portuguese diplomatic agents to acquire globes, telescopes, quadrants, sextants and a remarkable map collection from some of the best makers in Europe for his Palace of Ribeira, Lisbon.[6] The palace library was organised so that scientific instruments could be used according to the latest theoretical treatises (*Fig. 2*). A considerable collection of clocks, many made by the finest clockmakers, was also assembled in the main tower of the Palace, near the king's private rooms and Library.[7] João V also financed an astronomical observatory in the Palace. It was organised by Giovanni Battista Carbone (1694–1750) and Domenico Capassi (1694–1736), both mathematicians at the Jesuit College of Santo Antão, Lisbon.[8] The first astronomical observations were made as early as 1722.[9] Very few instruments survived the devastating 1755 earthquake and subsequent fire that destroyed the Palace.[10]

Apart from the palace collections, João V also sponsored important scientific institutions in Portugal (*Fig. 3*). In Lisbon, in the 1730s, he sponsored the construction of an astronomical observatory at the Jesuit College of Santo Antão, at the time the most important astronomy and natural philosophy teaching institution in Portugal. He also provided the College with a fine scientific library, scientific equipment and teachers.[11] Moreover, in 1747, João V allocated part of his Palace of Necessidades, also

---

6 About the Palace of Ribeira, see Nuno Senos, *O Paço da Ribeira (1501–1581)*, Lisbon, 2002; Bruno A. Martinho, *O Paço da Ribeira nas vésperas do Terramoto*, Master dissertation in History of Art, Universidade Nova de Lisboa, 2009.

7 About the collections of João V at Ribeira, see Angela Delaforce, *Art and Patronage in Eighteenth Century Portugal*, London, 2002, pp. 67–117; Marie Thérèse Mandroux França, Maxime Préau (eds.), *Catalogues de la collection d'estampes de Jean V, roi du Portugal*, Lisbon-Paris, 2003, 3 vols.; de Carvalho, *A Física Experimental...*, cit. (n. 5), pp. 54–56; Lilia M. Schwarcz, *A longa viagem da biblioteca dos reis. Do Terramoto de Lisboa à Independência do Brasil*, Lisbon, 2007, pp. 71–80.

8 Luis Tirapicos, *O telescópio astronómico em Portugal no século XVIII*, Master dissertation in the History of Science, University of Lisbon, 2010, pp. 25–32 and 124–125.

9 João Baptista Carbone, "Observationes astronomicae habitae Ulyssipone, Anno 1725, & Sub Init. 1726 [...]", *Philosophical Transactions* 34 (1726–1727), pp. 90–101. See Tirapicos, op. cit. (n. 8), pp. 25–32.

10 We know of at least four: two Coronelli globes, acquired in 1723 from Johannes Schalbruch, professor at the University of Amsterdam; a magnet assembled in Lisbon by William Dugood, member of the Royal Society; and a telescope lens made by Antonio Degola for Francesco Bianchini (1662–1729). About the latter, see Luis Tirapicos, Gilberto Pereira, "A Rare Telescope Objective Lens by Antonio Degola", *SIS Bulletin* 112 (2012), p. 40.

11 See Henrique Leitão, *A Ciência na "Aula da Esfera" no Colégio de Santo Antão, 1590–1759*, Lisbon, 2007. A survey of the scientific equipment of the College in 1725 is in Carbone, "Observationes astronomicae...", cit. (n. 9), pp. 90–101.

Fig. 2. Engraving by Pedro de Rocheford, Lisbon, 1732, from Luís Caetano de Lima, *Geographia Histórica de Todos os Estados Soberanos da Europa*, Lisbon, 1734–36, vol. 1, dedicated and offered to King João V by the author (National Library of Portugal, Lisbon, BNP CA. 614P).

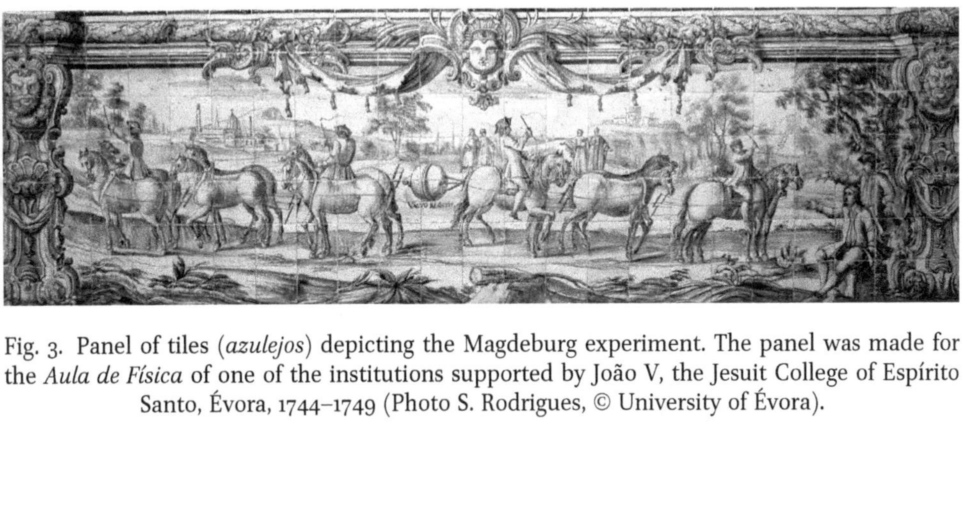

Fig. 3. Panel of tiles (*azulejos*) depicting the Magdeburg experiment. The panel was made for the *Aula de Física* of one of the institutions supported by João V, the Jesuit College of Espírito Santo, Évora, 1744–1749 (Photo S. Rodrigues, © University of Évora).

in Lisbon, to the Congregation of the Oratory, together with a cabinet of physics, a library and an annual stipend (*Fig. 4*).[12]

In 1751, the Oratorians initiated teaching in the sciences and humanities at the Palace of Necessidades. The library and instrument cabinet were regularly used from 1754 for public demonstrations, mostly in optics and electricity. The royal family and the court often attended these. In 1750, after the death of King João, his son José I (1714–1777) continued to visit the Oratorian Cabinet.[13]

The Late 1700s: The Education and Leisure of Princes José and João and the Ajuda Cabinet of Physics

José I, and his daughter Maria I (1734–1816), followed King João's tradition and promoted the organisation and development of scientific collections. These were assembled at the Palace of Ajuda, the main Lisbon residence of the royal family after the 1755 earthquake.[14] Physicists, mathematicians and naturalists were called from all over Europe to administer the new collections and, in some cases, provide public access. Aside from their educational purpose, these collections also enabled the display of wealth, power and social prestige, as well as an "enlightened", "modern" belief in scientific development.

In the 1760s, a museum of natural history, a botanical garden, a chemistry laboratory, a library and a drawing and engraving course were established in Ajuda, initially under the direction of the Italian naturalist Domenico Vandelli (1735–1816). The natural history collections greatly benefited from specimens collected in Brazil, until 1821 a Portuguese colony. Although initially dedicated to the instruction and leisure of young Princes José and

---

[12] Francisco C. Domingues, *Ilustração e Catolicismo. Teodoro de Almeida*, Lisbon, 1994, pp. 28–29 and 74–78; Manuel H. Corte-Real, *O Palácio das Necessidades*, Lisbon, 1983, pp. 13–18; Leonor Ferrão, *A Real Obra de Nossa Senhora das Necessidades*, Lisbon, 1994, pp. 178–180; de Carvalho, *A Física Experimental...*, cit. (n. 5), pp. 50–61, 71–72 and 84.

[13] An anonymous friend of Father Teodoro de Almeida (1722–1804), described a King's visit to the Cabinet as follows: "The King often operated the machines with his royal hands, feeling, full of curiosity, the experience of the natural phenomena that were described to him"; *Vida do Padre Teodoro de Almeida*, 1830: Lisbon, Arquivo National da Torre do Tombo (ANTT), Manuscritos da Livraria, no. 2316.

[14] About the Palace of Ajuda, see Maria Isabel B. Abecasis, *A Real Barraca. A Residência na Ajuda dos Reis de Portugal após o Terramoto (1756–1794)*, Lisbon, 2009.

Fig. 4. Engraving by Caetano, Oficina Régia Lithographica, Lisbon, 1824, representing a planetarium built by Teodoro de Almeida (1722–1804), professor at the Congregation of the Oratory in 1796. The planetarium itself has been lost, but its wooden base is presently at the Museum of the University of Lisbon (National Library of Portugal, Lisbon, BNP E. 1126V).

João, two sons of Queen Maria I, the Ajuda Natural History Museum and Botanical Garden opened to the public in 1768.[15]

It was also in the 1760s that a cabinet of physics was organised at Ajuda for the education of José and João. According to Manuel do Cenáculo (1724–1814), one of the princes' first tutors, their education should be informal, useful and experimental. Apart from history, philosophy and literature, the princes studied geography, geometry, and physics.[16]

Little is known about the cabinet's early years. In 1766, Martinho de Melo e Castro (1716–1795), the Portuguese ambassador in London, dispatched to Lisbon an optical chamber with 187 engravings, possibly for the cabinet.[17] It was only in 1777, with the appointment of the Venetian mathematician Michele Franzini (1740?–1810) as both head of the cabinet and mathematics teacher to the two princes, that the cabinet began to show considerable development.[18] Franzini started with books. In 1778 he acquired the works of Nollet, Sigaud de la Fond, Para du Phanjas, 's Gravesande and Desaguliers, among others.[19] He organised the cabinet in accordance with eighteenth-century physics and practical mathematics, therefore integrating astronomy, geography, navigation, surveying, military architecture and strategy. In terms of diversity, size and typology,

---

[15] Carlos Almaça, *Museus de zoologia e investigação científica*, Lisbon, 1985; Almaça, *Bosquejo histórico da zoologia em Portugal*, Lisbon, 1993; João Carlos Brigola, *Colecções, Gabinetes e Museus em Portugal no século XVIII*, Lisbon, 2003.

[16] Manuel do Cenáculo, "Diário", 1766–1780: Évora, Biblioteca Pública de Évora (BPE), CXXIX/I-17, fls. 231–237. About Cenáculo's teaching activity at the court, see Ferro, op. cit. (n. 3), pp. 50–55; Jean Marcadé, *Frei Manuel do Cenáculo Vilas Boas. Évêque de Beja, Archevèque d'Évora (1770–1814)*, Paris, 1978, pp. 62–66.

[17] Martinho de Melo e Castro, Letter to Diogo Mendonça Corte Real, 16 November 1766: Lisbon, ANTT, Conselho de Guerra, Mç. 297. Also cited in Ferro, op. cit. (n. 3), p. 58; de Carvalho, *A Física Experimental*..., cit. (n. 5), pp. 81–82.

[18] Franzini first came to Lisbon to teach at the Colégio dos Nobres (College of the Nobles), a teaching institution for Portuguese aristocracy created in Lisbon in 1761. About the College, see de Carvalho, *História da Fundação do Colégio Real dos Nobres*..., cit. (n. 5). Also in the 1760s, Giovanni Dalla Bella (1730–1823) was organising the Cabinet of Physics of the Colégio dos Nobres. In 1773, this Cabinet was transferred to the University of Coimbra, where Dalla Bella continued to teach until he returned to Padua in 1790. Although it is likely that Franzini and Dalla Bella exchanged ideas about instruments and makers (both taught at the Colégio dos Nobres and the University of Coimbra), we have found no concrete evidence and their relationship requires further research. A considerable number of instruments from the original Dalla Bella Cabinet still exists at the Science Museum of the University of Coimbra; see Maria C. Ruivo (ed.), *Engenho e a Arte. Colecção de Instrumentos do Real Gabinete de Física*, Lisbon, 1997.

[19] Expenses from 1778 and 1779: Lisbon, ANTT, Casa Real, Cx. 3113 and 3117.

the Ajuda Cabinet had many aspects in common with other royal cabinets in Europe.[20]

The cabinet was significantly developed between 1777 and 1794. Franzini continuously acquired scientific instruments and books, mostly through the instrument maker John Hyacinth de Magellan (1722–1790) and Rodrigo de Sousa Coutinho (1755–1812), the Portuguese Ambassador in Turin. For more than 15 years, and following Franzini's instructions, Magellan and Coutinho continuously acquired instruments from across Europe for the cabinet.

Magellan was well connected with many scientists and instrument makers in Europe. He lived in London and started buying instruments for the Portuguese Court in the late 1770s. In 1777, Magellan acquired a considerable number of optical and astronomical instruments to be used in border demarcations in Brazil. In 1780 and 1781, he also commissioned instruments for the University of Coimbra and the Academy of Sciences in Lisbon.[21] One of Magellan's most important acquisitions for the Ajuda Cabinet was made at the auction of the Duke of Lorraine's collections in Brussels in 1781. He apparently acquired 16 boxes of instruments. Although a detailed list remains to be found,[22] a collection of medals, a collection of roman coins and a chronometer were among the acquisitions made by Magellan at the auction.[23] In June 1785, he also dispatched to Lisbon a pedometer, an achromatic ocular lens and a bubble level made in Venice.[24]

Two years later, Magellan made new acquisitions in London, again under Franzini's instructions. Twelve boxes from George Adams arrived at

---

20 Peter de Clercq, "Science at Court: The Eighteenth Century Cabinet of Scientific Instruments and Models of the Dutch Stadholders", *Annals of Science* 45 (1988), pp. 113–152; Pascale Mormiche, "Instruments et pratique scientifique: L'évolution des choix et des pratiques dans le Cabinet de physique des enfants de France à Versailles d'après les inventaires des Menus-Plaisirs (1758–1765)", *Bulletin du Centre de Recherche du Château de Versailles* (2011), on-line publication http://crcv.revues.org/11584 (accessed November 2012); Lewis Pyenson, Jean-François Gauvin (eds.), *L'Art d'Enseigner la Physique. Les appareils de démonstration de Jean-Antoine Nollet (1700–1707)*, Sillery (Québec), 2002.

21 About Magellan, see Isabel M. Malaquias, *A obra de João Jacinto Magalhães no contexto da ciência do século XVIII*, PhD thesis in Physics, University of Aveiro, 1994.

22 Victor Rasquin, *Les instruments scientifiques dans les collections de Charles de Lorraine, édition annotée de l'inventaire après décès et du catalogue de vente*, Bruxelles, 2002.

23 John H. Magellan, Letters to Aires Sá e Melo, 27 August 1781, 17 April 1782 and 20 April 1784: Lisbon, ANTT, MNE, Cx. 9, Mç. 12; also cited in de Carvalho, *A Física Experimental...*, cit. (n. 5), pp. 81–82.

24 João Antonio Pinto da Silva, "Livros e mais cousas providas pera uso de SS. Altezas o Príncipe Nosso Senhor", 26 June 1785: Lisbon, ANTT, Casa Real, Cx. 3139.

Ajuda in 1787. The instruments were of the finest English craftsmanship, designed by Edward Nairne, George Adams and Nicholson.[25] Among these were a Heron's fountain, an electrical machine and apparatus, a Papin's boiler, an Atwood Machine, an air pump, a eudiometer and a reflecting telescope by Nairne.[26] The Ajuda Cabinet would also include instruments by Ramsden and Dollond. Magellan was also a renowned instrument designer and he may have sold or offered his own instruments to Ajuda.

Rodrigo de Sousa Coutinho was a scholar with interests in the sciences, arts and humanities. He had a profound interest in the economic and cultural development of Portugal. Also well connected among European elites, Coutinho maintained close relations with Correia da Serra (1750–1823)[27] and Alessandro Volta (1745–1827), among others.[28] He played a crucial role in the development of the Ajuda Cabinet, enlarging its scope and aims.

From 1779 to 1789, Coutinho ordered 72 models and machines for the instruction of the two princes. Initially, he gave particular attention to military strategy and ballistics. The protection and fortification of ports and borders were crucial for national defence (Portugal had suffered a Spanish invasion in 1762),[29] but also to ensure the continuity of secure commerce and trade. He commissioned 27 replicas of fortification and artillery models from the Regie Scuole Teoriche e Pratiche di Artiglieria e Fortificazione in Turin. Earlier, José Joaquim de Miranda Rebelo, Coutinho's secretary, had visited the Turin Royal Schools and was impressed by their collection of teaching models. He shared his admiration with Coutinho,

---

[25] Presumably William Nicholson (1753–1815); Paolo Brenni, personal communication, 26 September 2012.

[26] Magellan, "General account of the articles contained in each case", 23 October 1787: Lisbon, ANTT, MNE, Cx. 707.

[27] Correia da Serra and Sousa Coutinho mantained an intense epistolary exchange. Correia da Serra acted as a scientific agent to Countinho, providing him with detailed information on scientific innovations from Paris, London and the United States, where he lived. See Ana Simões, Maria Paula Diogo, Ana Carneiro, *Cidadão do Mundo: Uma biografia científica do Abade Correia da Serra*, Porto, 2008, pp. 69–75.

[28] Besides Volta, Sousa Coutinho got acquainted with other Italian scientists during his embassy in Torino until 1794, e.g. Carlo Antonio Napione (1756–1814) and Filippo Nicolis di Robiland (1723–1783). For a biography of Sousa Coutinho, see Andrée Mansuy-Diniz Silva, *Portrait d'un homme d'état: D. Rodrigo de Sousa Coutinho (1755–1812)*, Lisbon, 2002–2006, 2 vols; Júlio J. Rodrigues da Silva, *Ideário político de uma elite de estado. Corpo diplomático (1777–1793)*, Lisbon, 2002, pp. 505–636 and 1011–1104.

[29] Joaquim V. Serrão, *História de Portugal*, Lisbon, 1977–2000, 18 vols.; v. 11, pp. 58–63; Monteiro, op. cit. (n. 3), pp. 152–164.

who ordered replicas of all models for Ajuda.[30] Each model represented a technique of military architecture, depicted in the course *Dell'architettura militare per le regie scuole teoriche d'artiglieria* (1778), by Alessandro Papacino D'Antoni (1714–1786), Director of the Schools.[31] From 1779 to 1783, the replicas were made and sent to Ajuda (*Fig. 5*).[32]

Coutinho also commissioned for Ajuda 20 miniature models to illustrate the manufacture and power of gunpowder from a treatise by D'Antoni.[33]

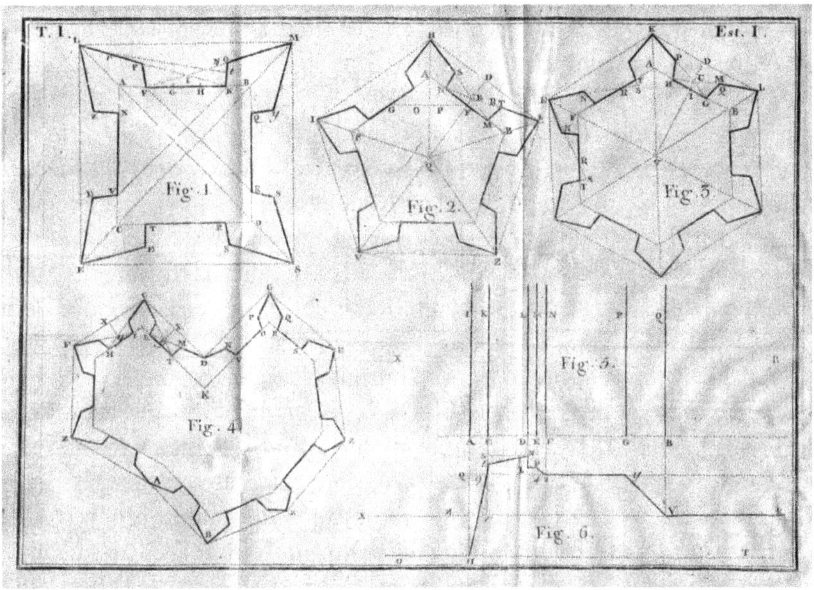

Fig. 5. Engraving from Alessandro Papacino D'Antoni, *Dell'architettura militare per le regie scuole teoriche d'artiglieria*, Torino, 1778; Portuguese edition of 1790, translated by Matias José Dias Azevedo, Lisbon, vol. 1, pl. 1 (National Library of Portugal, Lisbon, BNP ENC. 6).

[30] José J. Miranda Rebelo, Letter to Rodrigo de Sousa Coutinho, 29 September 1780: Lisbon, ANTT, Casa de Linhares, 64/4.

[31] Alessandro V. Papacino D'Antoni, *Dell'architettura militare per le regie scuole teoriche d'artiglieria*, Torino, 1778.

[32] Rodrigo de Sousa Coutinho, *Memória dos Modelos de Fortificação... que se tem remetido e vão a expedirse para o serviço de S. A. Real o Príncipe*, 1782–1783: Lisbon, ANTT, MNE, Cx. 861; de Sousa Coutinho, *Memória dos Modelos de Fortificação... que se tem remetido e vão a expedirse para o serviço de S. A. Real o Príncipe*, 1785: Ibid., Cx. 862; de Sousa Coutinho, *Memória dos Modelos de Fortificação... que se tem remetido e vão a expedirse para o serviço de S. A. Real o Príncipe*, 1788: Ibid., Cx. 864.

[33] A.V. Papacino D'Antoni, *Esame della Polvere dedicato a Sua Sacra Real Maestà*, Torino, 1765.

By the end of 1788, the cabinet had received 39 fortification models and 10 artillery models from Turin, all made by two craftsmen – a carpenter and a locksmith – from the Royal Schools under the supervision of D'Antoni himself.[34] The cabinet also had several miniatures and models of Portuguese military and trade boats. In 1786, a collection of 40 tools and instruments used in arts and crafts was also included.[35]

In his interactions with Franzini, Coutinho insisted on science and technology as catalysts for Portuguese economic development. In his view, the future King of Portugal should learn about technological and industrial innovation. Coutinho demonstrated a special interest in the *alla piemontese* technique of spinning silk, praised by the *Encyclopédie* as the best invention of the eighteenth century.[36] In February 1783, he ordered five models of silk mills and machines used in Piedmont for the Ajuda Cabinet. This had concrete economical impact, because the first silk-spinning factory operating *alla piemontese* opened in Chacim, northern Portugal, less than five years later.[37] Coutinho also sent to the cabinet several models of innovative hydraulic engines, such as the famous Marly Machine that brought water from Louveciennes to Versailles, models of canals and so on.[38]

It is likely that Vicente de Sousa Coutinho (1726–1792), uncle of Rodrigo and ambassador in Paris, also played a relevant role in the acquisition of instruments. Although his role in the Ajuda Cabinet needs further research, he was also a scholar and well connected in Versailles. He knew Lalande, Laplace, d'Alembert and l'Abbé Bossut, among others.[39] It is likely that he was responsible for the acquisition from French makers of

---

34 Miranda Rebelo, Letter to Rodrigo de Sousa Coutinho, 7 August 1782: Lisbon, ANTT, MNE, Cx. 861.

35 Expenses from July 1782: Ibid., Casa Real, Cx. 3121; Expenses from January and February 1786: Ibid., Cx. 3143.

36 de Sousa Coutinho, *Reflexões políticas sobre os Meios de Estabelecer em Portugal a Cultura e Manufactura da Seda*, 20 March 1784, and *Plano de Lotaria sobre os Meios para Fundar o Estabelecimento da Seda em Trás-os-Montes*, 25 October 1786: Ibid., Cx. 862.

37 About Chacim, see José M.A. Mendes, *Trás-os-Montes nos finais do Século XVIII: Alguns aspectos económico-sociais. Segundo um manuscrito de 1796*, Lisbon, 1995; Francisco A. Lourenço Vaz, *Instrução e Economia. As Ideias Económicas no discurso da Ilustração portuguesa (1746–1820)*, Lisbon, 2002, pp. 409–428.

38 de Sousa Coutinho, Letter to Luis Pinto de Sousa, 19 March 1783: Lisbon, ANTT, MNE, Cx. 861; de Sousa Coutinho, Letter to Luis Pinto de Sousa, 28 June 1784: Ibid., Cx. 862; de Sousa Coutinho, Letter to Luis Pinto de Sousa, January 1787: Ibid., Cx. 864.

39 Mansuy-Diniz Silva, op. cit. (n. 28), p. 80.

two silver compasses,[40] a gold mathematical *necessaire*,[41] and an orrery by the Dumotiez brothers.[42]

Like many late-eighteenth-century and nineteenth-century cabinets of physics, the instruments at Ajuda were organised typologically and were used. A visitor to the cabinet, Marc-Marie de Bombelles (1744–1822), seems to describe four independent rooms: one for instruments, another for models (military fortifications and tools for arts and crafts), a third with artillery models and a fourth with hydraulic machines.[43]

Books seem to have played an important role complementing teaching and experimenting. Apart from the books mentioned earlier, between 1777 and 1794 Franzini bought 190 more books for the cabinet. The largest subjects represented (c. 40%) were economic theory – including the works of Jacques Necker and Adam Smith – industrial and agricultural development, and arts and crafts. Geography and natural history were also included in the cabinet's library. Military and naval strategy occupied a significant place as well, especially the works by architect Sébastien Vauban and treatises on the Prussian army and the military achievements of Frederick the Great. Mathematics and physics were well represented too, with works by Étienne Bezout, Jacques Antoine Cousin, Pieter van Musschenbroek and Joseph-Louis Lagrange. Franzini also assembled a considerable collection of maps, including the most recent publications of the time, such as the *Atlas Russicus* (1745) and the *Atlante español* by Bernardo Espinalt (1778). Today the majority of these books are in the national libraries of Lisbon and Rio de Janeiro.

By the end of 1793, Franzini left for Venice. He was replaced by Joan Joseph Solner, a Dutch physicist who lived in Lisbon.[44] Solner initiated the cabinet's catalogue but this document remains to be found. However, in

---

40 Expenses from April 1779: Lisbon, ANTT, Casa Real, Cx. 3112.

41 Expenses from April 1778: Ibid., Cx. 3589.

42 No expense records exist, but this French orrery is mentioned in Tibère Blanc, *L'Etat des Instrumens de Physique experimentale, Modeles de Fortifications et des Arts et Métiers*, 28 July 1808: Paris, Archives Nationales de France (ANF), Carton AJ/15/600. Therefore we suggest it may have been acquired via Vicente de Sousa Coutinho.

43 Marquis De Bombelles, *Journal d'un Ambassadeur de France au Portugal. 1786–1788*, Paris, 1979, pp. 271–272. See João C. Brigola, *Os viajantes e o livro dos museus. As colecções portuguesas através do olhar dos viajantes estrangeiros (1700–1900)*, Porto, 2010. The Cabinet space still exists today, although the Palace of Ajuda has been largely rebuilt in the nineteenth century.

44 Solner was well known in Lisbon for his public demonstrations of physics. In 1794, he sold more than 300 instruments and machines to the Royal Academy of Sciences. See de Carvalho, *A Física Experimental*..., cit. (n. 5), p. 88.

July 1808, following Napoleon's occupation of Lisbon, Tibère Blanc (1765–1812), a member of the French Consulate, made a catalogue of the cabinet.[45] Although at the time Prince José had deceased (he died at twenty-eight, of smallpox) and João was already forty, the cabinet catalogued by Blanc was likely to be similar to the one organised by Franzini.

In 1808 it comprised 324 instruments, models and artefacts, covering the areas of hydraulics, physics, mathematics, military strategy, ballistics and military construction, among others. According to the results of the survey, didactic instruments were plentiful and diverse: 50 instruments to illustrate the principles and laws of "*Pneumatique*" and "*Magnétisme*"; 39 items of electrical apparatus and 23 optical instruments, including four telescopes and one curved mirror. Measuring instruments were also abundant: thermometers and hygrometers, balances, rulers and compasses. Astronomy occupied an important role also: apart from globes and telescopes, there were armillary spheres and orreries, beside other instruments for "*Optique*". Laws of "*Méchanique*" were exemplified by several models and instruments. "*Hydrostatique*" was represented by several pieces of apparatus and models, including one Heron's fountain, suction pumps and one vacuum pump. Finally, Blanc's catalogue also lists the 72 Turin models mentioned earlier (by then almost twenty years old).

Little is still known about the cabinet's daily use during the Franzini years. The princes seem to have had a predilection for the cabinet. In 1788, when Prince José met with the English aristocrat William Thomas Beckford (1760–1844), one of the first questions he asked him was if Beckford had visited "his" cabinet of physics.[46] Prince João also used the cabinet regularly, if not daily, to receive lessons and attend demonstrations. On other occasions, instruments, machines and models were taken to the princes' private rooms (*Fig. 6*).[47] Frequent instrument repairs also confirm intense use.[48]

The cabinet's existence and use seem to have coincided with the youth and education of the two princes. Sources suggest limited use, if any, by other generations of the royal family. After 1795, when Prince João was eighteen, acquisitions became increasingly scarce and the cabinet gradually

---

[45] Blanc, op. cit. (n. 42).

[46] William T. Beckford, *Italy with sketches of Spain and Portugal*, London, 1834, 2 vols.; v. 2, p. 223. Unlike the Ajuda Museum and Botanical Garden, the Cabinet of Physics was not open to the public, and only guests were allowed.

[47] Joan J. Solner, Letter to Marquês de Tancos, 17 January 1794: Lisbon, BNP, Cod. 9815.

[48] Expenses from 1779, 1780, 1787 and 1789: Lisbon, ANTT, Casa Real, Cx. 3112, 3116, 3123 and 3146.

Fig. 6. Portrait of Prince José, aged twelve, with a terrestrial globe. Oil on canvas by Miguel António do Amaral, Lisbon, c. 1773, State Hermitage Museum, St. Petersburg, Inv. GE-4430 (photo by V. Terebenin, L. Kheifets, Y. Molodkovets, © The State Hermitage Museum).

declined. Moreover, in 1807, the entry of Napoleon's armies resulted in the transfer of the court and capital of the Portuguese Empire to Rio de Janeiro. From Rio, Prince João ordered the shipment of the Ajuda library and part of the physics cabinet, most likely to continue the education of his son Pedro (1798–1834), who was then nine years old.[49] Pedro would later become the first Emperor of Brazil. After the court's return to Lisbon in 1821, what was left of the cabinet in Lisbon was gradually abandoned. In 1854, half of the items listed in Blanc's 1808 catalogue no longer existed.[50]

Concluding Remarks

Princes José and João's cabinet of physics in late-eighteenth-century Lisbon provides an interesting example for the history of royal collections and, more generally, for the history of elite education in peripheral European countries. It shared many characteristics with other cabinets of physics, for example in universities and academies: its size was average, instruments were in intense use, access was limited, location was reserved, and it was complemented by a library. The Ajuda Cabinet had many instruments of science, as we would call them today, but it also had a strong focus on state-of-the-art industrial, military and technological innovations, aimed at training future statesmen of a late-eighteenth-century empire that was rich in resources yet poor and underdeveloped. All instruments had to be acquired abroad. This combination of purposes is the result of the personality and drive of two men – Magellan, more science-oriented, and Coutinho, a diplomat interested in industrial and economic development – supervised by the Italian mathematician Michele Franzini. This combination also possibly explains the cabinet's "frugality", namely the limited number of highly crafted and ornamental instruments that were common in other royal cabinets in Europe, such as the ones in London, Dresden and Prague.

What happened to these instruments? Certainly, many were lost and dispersed due to wars, moves, auctions, revolutions and natural abandonment. There were other royal cabinets of physics in the nineteenth century in Lisbon and Rio. In 1910, when monarchy was abolished in Por-

---

[49] *Colecção das Leis do Brasil de 1810*, Rio de Janeiro, 1891, p. 220; Expenses from 1809 to 1812: Lisbon, ANTT, Casa Real, Cx. 3245 and 3255.

[50] *Relação dos objectos existentes nas sallas do Antigo Gabinete de Physica do Real Palácio d'Ajuda*, 14 June 1854: Ibid., Cx. 4135.

tugal, there were about 1,000 scientific instruments in the two main royal palaces in Lisbon: Ajuda and Necessidades.[51] Today we have identified about 200 instruments, from the sixteenth to the late-nineteenth century, with confirmed Portuguese royal provenance. They are dispersed in palaces, museums and schools in Portugal and Brazil (*Pl. XI*). A considerable number still exists at the Palace of Ajuda. Further research into the dispersals still needs to be done, combined with material study of the surviving instruments.

The study of cabinets of physics can provide important insights into the development of experimental inquiry, theoretical conjectures, research and teaching practices, technological innovation and trade. Cabinets and instruments are increasingly being considered as sources for the history of science and technology. Among cabinets of physics, royal cabinets of physics are a special case due to the wealth and power of the owners. Wealth and power may bring a certain "antiquarian" perspective – possibly the most beautiful European instruments are from royal collections – but it can also bring broader historical, economic, social and political contexts. The history of royal collections and libraries often materialises, in many aspects, the history of countries themselves.

51 *Inventário Judicial do Palácio da Ajuda*, 1911–1913: Lisbon, Palácio Nacional da Ajuda (PNA) – Archive; *Inventário Judicial do Palácio das Necessidades*, 1910–1926: Ibid.

THE GAZOLA FAMILY'S SCIENTIFIC CABINET: POLITICS, SOCIETY AND SCIENTIFIC COLLECTING IN THE TWILIGHT OF THE REPUBLIC OF VENICE

Ivano Dal Prete*

Introduction

In this paper I will outline the history and composition of the scientific instruments cabinet assembled in the late 1700s by an aristocratic family of the Venetian hinterland, the Gazolas from Verona. This family has long been known to scholars of early-modern natural history, thanks to their museum of fossil fish that in the mid-1790s was arguably the richest of its kind in Europe. Because of its renown and scientific importance, the collection was moved in 1796 to the Muséum d'Histoire Naturelle in Paris. The Gazola collection of scientific instruments, on the contrary, was virtually unknown until recently, when I had the opportunity to survey the surviving items for the first time (they are now part of the collection of the Liceo "Maffei" in Verona) and to compile a reasonably accurate catalogue of its original composition. Gazola's cabinet was one of the most complete private collections of scientific instruments in the eighteenth-century Republic of Venice, including apparatus that spanned across almost every field of experimental philosophy. In the following pages, I will describe some of the most notable instruments among those I could identify, and I will provide in the appendix a list of the items that belonged in the physics cabinet. I will also explore the history of this collection, focusing on its cultural context as well as on the Gazolas themselves, in order to shed light on the social and political issues that in the late eighteenth century led to the rise of aristocratic scientific cabinets in the Republic of Venice.

With their vast naturalistic and experimental collections, the Gazolas were the most conspicuous local interpreters of a widespread trend: Krzysztof Pomian noted that natural philosophy and especially natural history enjoyed a growing presence in Parisian cabinets in the eighteenth

* I am grateful to Giorgio Modena and Giorgia Vettori (Liceo Classico "Scipione Maffei", Verona) for their assistance and encouragement; to Sofia Talas, Anita McConnell and Michel Morizet for their suggestions and help; to Paola, as always, for patience and love.

century: "After 1720", Pomian argued, "and especially after 1730, Parisian collectors began to forsake medals . . . Their attention was attracted by natural history".[1] The composition of aristocratic cabinets in Europe, as well as in Venice and the Venetian hinterland, followed a similar trend, which entailed an increased role in patrician collections for scientific instruments and cabinets of experimental philosophy. Recent scholarship has shown, however, that it would be naive to assume that the geographical and cultural peripheries of Enlightenment Europe simply followed the trends and the models of the capitals; as the knowledge of local contexts improves, the focus has increasingly shifted to the diversity of eighteenth-century Europe and to how local agendas shaped scientific culture and practices.[2] The spread of Enlightenment values and ideals, the new appreciation of a rational understanding of nature, mediated by Newtonian natural philosophy, and the emulation of the great capitals (Paris in particular) were certainly elements that contributed to the spread of scientific culture and objects among provincial aristocracies. I would argue, however, that in the Venetian context this phenomenon was above all a reaction to the potential social and political impact of "modern" natural philosophy, rather than an instance of its reception. By appropriating recent scientific practices, methods and instruments, local elites intended to take part in the elaboration of natural knowledge in order to employ it in the service of conservative political agendas; their final aim was the preservation of the ideological and institutional foundations of *ancient régime* society.

Collecting and Natural Philosophy in the Eighteenth-Century Republic of Venice

The Gazolas did not belong to the most ancient and prestigious families of the Venetian hinterland. In the family archive and in the fiscal records they appear as landowners and "citizens" from at least the late 1500s,[3] but they managed to be ennobled only in the 1730s.[4] They had to wait until

---

[1] Kristofsz Pomian, *Collezionisti, amatori e curiosi. Parigi-Venezia XVI–XVIII secolo*, Milan, 1989, p. 164.

[2] See, among recent literature: Kostas Gavroglu, Manolis Patiniotis, *et Alii*, "Science and Technology in the European Periphery: some Historiographical Reflections", *History of Science* 46 (2008), pp. 153–175; Charles W.J. Withers, *Placing the Enlightenment. Thinking Geographically about the Age of Reason*, Chicago-London, 2007.

[3] Verona, Archivio di Stato (ASVr), Archivio Gazola, n. 48.

[4] Andrea Cartolari, *Famiglie già ascritte al nobile consiglio di Verona*, Verona, 1854, part 2, p. vii.

after the fall of the Republic of Venice in order to gain full access to the councils that administered Verona (the second largest city in the state), traditionally reserved to the most illustrious patrician houses. On the other hand, the Gazolas were well known for their traditions in medicine and natural history. In the last years of the seventeenth century Giuseppe Gazola – one of the most celebrated "neoteric" physicians of the time – led a physico-medical academy that pitted the new mechanistic medicine against the local medical college, loyal to the Galenic tradition and habitually reserved for members of the aristocracy.[5]

Ennobled and married into the nobility after the 1730s, but still seen as *parvenus* within the local elite, the Gazolas consistently sought to increase their prestige and visibility through their technical and scientific culture. Their choice may appear unusual, since in the early eighteenth century the role of natural philosophy in Italian aristocratic culture and education was often negligible compared to the primacy of poetry, music and painting. A description of the largest Veronese museums published in 1732 by Marquis Scipione Maffei (one of the most influential Italian intellectuals) mentions above all paintings, medals, antiquities, marble busts, prints, manuscripts, etc. Such objects certainly formed the bulk of the aristocratic collections, including Maffei's own, which featured only "a few natural things".[6] Most patricians hardly received any mathematical or scientific education, the learning of a gentleman being measured in terms of his literary, antiquarian and musical culture.

Maffei himself, however, was convinced that contemporary natural philosophy should have an essential role in the renovation of the Italian urban aristocracies that was central to his political projects.[7] The Gazolas wholly endorsed his efforts and tried to match his ideal of patricians engaged in useful activities and in the promotion of the "common good". By the 1740s Count Andrea Gazola (Giuseppe Gazola's nephew) assembled a collection of scientific instruments, of which almost nothing is left and little more is known; several prints and drawings preserved in the City Library of Verona, however, show machines and experiments that he probably intended to publish in a work – tentatively called "Gazola's

[5] See: Ivano Dal Prete, *Scienza e Società nel Settecento Veneto*, Milano, 2008. Gazola's most influential work was Giuseppe Gazola, *Il mondo ingannato da' falsi medici e disingannato*, Venezia, 1716, reprinted at least a dozen times until the 1760s and translated into Dutch, French and Spanish.

[6] Scipione Maffei, *Verona Illustrata*, Verona, 1732, p. 396.

[7] See: Dal Prete, op. cit. (n. 5), part II. On Maffei's political ideas: Paolo Ulvioni, *Riformar il mondo. Il pensiero civile di Scipione Maffei*, Alessandria, 2008.

Fig. 1. The electric machine used by the Gazolas in 1747. The drawing is part of a set arguably composed for a work on "Gazola's Workshop" that was never published (Verona, Biblioteca Civica, *Laboratorio Gazoliano*, 1.9.n).

Workshop" – which was never completed. Among them, he submitted to Giovanni Poleni the project of a pile driver for hydraulic works,[8] whose model is now preserved at the Museum for the History of Physics at the University of Padua.[9] When the craze for electricity sprang up in European salons in 1746–47, the Gazolas readily took advantage of their expertise and traditions to take a leading role in the electric performances held in town and bring their salon to the centre of the local cultural life. Count Gazola and his young wife Massimiliana acquired or built an electrical machine, provided with a rotating glass cylinder, and placed it in their "Casino" on the banks of river Adige. The "Casino" enclosed a vast botanical garden where they performed a whole range of experiments, witnessed by many onlookers (*Fig. 1*).[10] Their large collection of fossils benefited from the paleontological richness of the territory,[11] but it also related to the heated controversies on the age and history of the Earth and seconded Maffei's stance against the use of the Bible in natural matters.[12]

Maffei's efforts, however, failed to arouse a lasting interest for natural philosophy and scientific practices among his peers; as they gradually merged into the local aristocracy, the Gazolas themselves tended to adopt aristocratic values and interests. While some of their instruments probably remained in the family cabinets, their fossils were sold to Maffei's main collaborator, the French botanist and antiquarian François Séguier, who left Verona for Nîmes after the death of his patron in 1755.

The "abuses of reason"

The Gazolas turned again to natural philosophy twenty years later, within a general re-orientation of the cultural priorities of the ruling elites of the Venetian state. In the second half of the century, the diffusion of materialist philosophies of French origin was increasingly perceived as a danger to the social stability of the Italian society; attacks to the authority of the Scriptures and to that of its interpreters threatened institutions that were

---

[8] Andrea Gazola, Letter to Giovanni Poleni, 22 August, 1748: Verona, Biblioteca Civica (BCV), b. 217.

[9] I am grateful to Sofia Talas for pointing out to me the existence of this model.

[10] S. Maffei, *Della formazione de' fulmini*, Verona, 1747, pp. 146–147.

[11] See: Giovanni Serafino Volta, *Ittiolitologia veronese del Museo bozziano ora annesso a quello del conte Giovan-Battista Gazola e di altri gabinetti di fossili veronesi*, Verona, 1796, 2 vols.; v. 1, p. 57.

[12] Dal Prete, op. cit. (n. 5), pp. 263–276.

essential pillars of the *ancien régime*. Newtonian astronomy and natural philosophy came to be accepted by Venetian elites not as the triumph of rationality, but rather as a natural theology that could be pitted against Cartesian materialism and the "abuses of reason" to which it contributed.[13]

In the same years, a raising technocracy of professional scientists and engineers increasingly questioned the role of amateurs and "dilettanti" as legitimate contributors to scientific knowledge. This tendency was embodied in the graduates of the military school for the engineers of the Venetian army established in 1759, whose students – mainly recruited among the lower classes – hoped to play a leading role in the administration of the Venetian state.[14] The separation between laboratory and salons that they advocated (a crucial break from early-modern scientific practices) was strongly opposed by the landed aristocracy, who feared the possible loss of control over the scientific discourse, especially after 1789 when social revolution became a real possibility.

From the mid-1770s on, scientific collections became increasingly visible in aristocratic galleries: fossils, minerals, botanical specimens that were previously neglected became sought-after items. By the early 1790s, almost all of the collections of natural objects previously owned by pharmacists or clergymen had been acquired by the nobility; their urban palaces and country villas were now decorated with frescoes of naturalistic or scientific subjects, often displaying telescopes, experimental apparatus or allegories of Newtonian astronomy adapted to aristocratic values.[15] Now firmly established within the ruling elite, the Gazolas placed themselves at the forefront of this movement employing natural specimens and scientific instruments to support a conservative political agenda.

Count Giambattista and his two brothers started rebuilding the family collections in the early 1780s, investing huge sums in the acquisition of instruments and fossils – in particular, large and well-preserved fossil fishes.[16] The family purchased sites of paleontological interest and private museums previously owned by physicians and apothecaries, whose objects were often sold by the heirs soon after their death. In the 1790s the paleontological museum owned by the Gazolas was probably the most

[13] Ibid., p. 343.

[14] See: Calogero Farinella, *L'Accademia Repubblicana. La Società dei Quaranta e Anton Mario Lorgna*, Milano, 1993.

[15] Dal Prete, op. cit. (n. 5), pp. 423–430.

[16] Lorenzo Sorbini, "Le collezioni naturalistiche veronesi nell'800", in Ezio Vaccari (ed.), *Le scienze della terra nel Veneto dell'Ottocento*, Venezia, 1998; pp. 95–107: 103.

remarkable in Europe, with over 1,200 fossil fishes, many of which were large and exceptionally well-preserved.[17] The Gazola museum was crucial to the publication of the *Veronese Ichthyology* (1796–1806), a lavish, large-format reproduction of the most important fossil fishes and plants preserved in local cabinets, funded and printed by a group of noblemen with the explicit aim of proving the truth of the Universal Flood against "atheists" and "pyrrhonists".[18]

Venetian aristocratic cabinets of scientific instruments should therefore be understood in the context of the conservative cultural and political agendas fostered by their owners. Not unlike fossils and natural specimens, instruments and experimental philosophy were employed in support of the preservation of cultural and religious beliefs deemed essential to the stability of Italian society.

Instruments and Politics in the Twilight of the ancien régime

The vast majority of the apparatus that comprised the Gazolas' cabinet in the late 1700s was certainly acquired between *ca.* 1780 and 1796, when the French invasion of the Republic of Venice and the ensuing turmoil prevented further acquisitions. A few of the instruments that I was able to identify, however, seem to have belonged to the family already in the 1740s. The cabinet was especially strong in pneumatics, "aerology",[19] electricity, optics, and chemistry. Aerology and chemistry became fashionable among the local elite after the first aerostatic flights: the Academy of Experimental Philosophy founded in Verona by Count Pietro Cossali in 1783 dealt in particular with the physics of the air, its properties, the relationship between air, light, sound, electricity, and the chemistry involved in the production of hydrogen.[20]

Giambattista Gazola demonstrated the political relevance of chemistry with his entry for a 1790 academic prize on the production of saltpeter:[21]

---

[17] Volta, op. cit. (n. 11), v. 1, pp. 68–69.

[18] Ibid., p. 283.

[19] In the late eighteenth-century and early nineteenth-century sources I consulted, "Aerology" was a discipline concerned with the physics of the atmosphere, in particular in relation to the new science of aerostatic flight. It seems to have included pneumatics (which deals with the mechanical characteristics of the air).

[20] BCV, "Cossali" Fund, ms. 1512. See also: ASVr, Archivio Dionisi-Piomarta, n. 630.

[21] Virgilio Giormani, "L'Accademia di Verona e il monopolio della fabbricazione del salnitro nella Repubblica Veneta", *Atti e Memorie dell'Accademia di Agricoltura Commercio ed Arti di Verona* 168–1 (1991–1992), pp. 129–154.

Gazola's new and "easy" method, which, according to the author, could be learned by every peasant and farmer, delivered a clear political message, as it entailed the abolition of the associated guild and, in general, the deregulation of the labor market and the weakening of the central government's control over the economy.[22]

The Gazola collection included an air pump "with Cavallo's improvement",[23] which is unlikely to have been made locally; the same holds true for some at least of the six barometers "of different provenance" and no doubt, for the most sophisticated pieces of Gazola's collection of optical instruments. While I could not find any evidence that Gazola ever engaged in systematic astronomical observations, his instruments could have equipped a complete and fully operational observatory. He might have been inspired by his co-citizen Antonio Cagnoli, a former pupil of the French astronomer Jérôme de Lalande who, in 1786, upon his return from Paris, erected in Verona a remarkable astronomical observatory with a collection of astronomical machines manufactured by Pierre Mégnié and Robert Robin.[24] Cagnoli's observatory was "comfortable, and not without elegance",[25] a space for both scientific investigation and display, visited by every learned citizen as well as by foreigners on their Grand Tour.

While Cagnoli relied on French makers, Gazola did not hesitate to purchase instruments from the most prestigious – and most expensive – British makers. His 2-ft astronomical quadrant was produced in Jesse Ramsden's workshop, arguably around 1790;[26] as it was customary in portable

[22] Giovanni Battista Gazola, *Sopra la facile produzione del nitro. Dialoghi due*, Verona, 1797.

[23] *Elenco delle macchine esistenti nel gabinetto fisico del Regio Liceo-Convitto di Verona*: Verona, Archives of the Liceo Classico "Scipione Maffei" (ALM), FAS 4 – 1811, n. 49. See: Tiberio Cavallo, "Description of an Improved Air-Pump, and the Account of Some Experiments Made with It", *Philosophical Transactions* 73 (1783), pp. 435–452.

[24] On Antonio Cagnoli's observatory and its instrumentation, see: Calogero Farinella, "Da Montesquieu a Lalande. Antonio Cagnoli e una specola privata del settecento", *Studi Settecenteschi* 17 (1997), pp. 227–264; Dal Prete, op. cit. (n. 5), pp. 430–441. See also: Derek Howse, "The Greenwich list of observatories: a world list of astronomical observatories, instruments and clocks, 1670–1850", *Journal for the History of Astronomy* 17 (1986), p. 41. Two of Cagnoli's instruments are described in Enrico Miotto, Giuseppe Tagliaferri, Pasquale Tucci, *La strumentazione nella storia dell'Osservatorio Astronomico di Brera*, Milano, 1989, pp. 47 and 61.

[25] Antonio Cagnoli, *Epoche principali della Vita di A. Cagnoli di Verona, ad opera del medesimo*, February 1798; Milan, Brera Observatory Archives, "Cagnoli" Fund.

[26] The quadrant features a pillar-construction frame as a means to increase rigidity with only a small weight penalty. Pillar-construction frames were used as early as 1781 by Edward Throughton, who patented his device in 1788. I owe this piece of information, and many others, to Anita McConnell.

astronomical instruments, it could also be used in surveying and terrestrial measurements by tilting its frame to a horizontal plane. While I did not find any documents concerning the purchase, the price list of Matthew Berge (who was Ramsden's successor at the head of the workshop) included astronomical quadrants with radii from 12 inches up and prices that ranged from 31 to 420 pounds (*Fig. 2*).[27] Gazola's optical cabinet included (among many other items) a transit instrument, achromatic telescopes of focal lengths up to 9 *piedi*,[28] and simple, compound, and solar microscopes. Some of the microscopes may have been purchased in Venice from Lorenzo Selva, who entertained close ties with the Veronese scientific milieu.[29] A Gregorian telescope manufactured in Paris by Le Bas *le jeune* may be one of the few surviving items from the early collections of the Gazolas, since Le Bas' activity is not documented beyond 1725. The instrument itself may be the oldest existing French Gregorian telescope, since it probably predates those that Paris & Conichon and Claude Passement started selling (according to the *Encyclopédie*) from 1733–1734 (*Fig. 3*).[30]

It is hardly surprising that electrical apparatus made up a substantial part of the Gazola's scientific cabinet. Even after the electrical craze of the late 1740s (when, according to Maffei, at least eight electric machines operated in town) electricity remained a subject of learned conversation; there is evidence that electric machines were occasionally dusted off and used for demonstrations and experiments in Veronese salons.[31] Electricity was probably the only subject in experimental philosophy the average

[27] Matthew Berge, *A Catalogue of Optical, Mathematical, and Philosophical Instruments Made and Sold by Matthew Berge, Successor to the Late J. Ramsden*, 1801: Winchester, Hampshire Record Office, 38M49/A4/7. I am grateful to Anita McConnell for letting me know of the existence of this catalogue, and for providing me with a copy. See also: Anita McConnell, *Jesse Ramsden (1735–1800). London's Leading Scientific Instrument Maker*, Ashgate, 2007.

[28] The Italian "*piede*" (pl. *piedi*), i.e. "foot", has been maintained to avoid confusion with the English length unit; *piede* could indeed refer to the Veronese foot (20.57 cm) or, most likely, to the French foot (32.48 cm).

[29] Dal Prete, op. cit. (n. 5), p. 338.

[30] Denis Diderot, Jean-Baptiste Le Rond d'Alembert, *Encyclopédie ou dictionnaire raisonné des sciences, des arts et des métiers*, Paris, 1751–1772, 35 vols.; v. 16, p. 44. I am grateful to Michel Morizet for sharing his competence on LeBas and early French Gregorian telescopes. On LeBas, see also: Maurice Daumas, *Scientific Instruments of the Seventeenth and Eighteenth Centuries* (translated and edited by Mary Holbrook), New York and Washington, 1972, pp. 73–74. On early French reflecting telescopes: Anthony Turner, "Claude Paris and the Early History of the Reflecting Telescope in France", *Journal of the Antique Telescope Society* 30 (2009), pp. 5–8.

[31] Dal Prete, op. cit. (n. 5), pp. 355–357.

Fig. 2. Portable quadrant by Ramsden, ca. 1790 (Verona, Liceo Classico "Scipione Maffei").

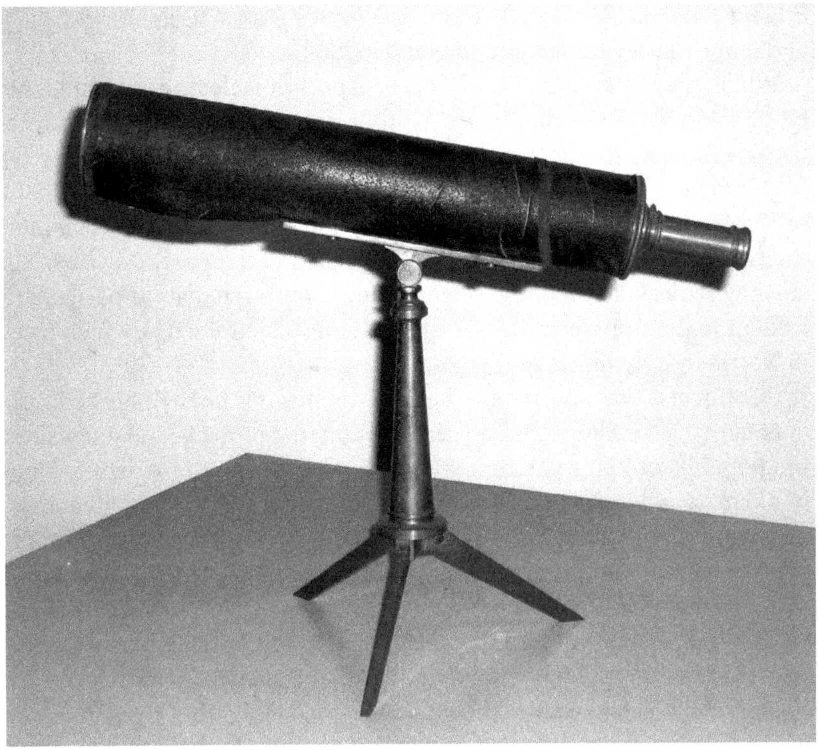

Fig. 3. Gregorian reflecting telescope by Le Bas, ca. 1725? (Verona, Liceo Classico "Scipione Maffei").

aristocrat had direct knowledge of even before the 1780s. In his 1779 *Instructive Dictionary for the Gentleman* (a sort of anti-*Encyclopédie*, meant to provide arguments for conversations against "modern philosophies"),[32] Count Antonio Montanari displayed a sketchy understanding of mathematics, chemistry and Newtonian astronomy, but he devoted 34 pages to the entry on "Electricity". He described electrical machines, lightning rods, the experiments he witnessed or actively took part in, showcasing an up-to-date knowledge of the literature on the topic and of on-going debates. An anonymous manuscript in the city library of Verona, composed around 1780 and largely based on Joseph Priestley's *History and*

[32] Antonio Montanari, *Dizionario Istruttivo per l'uomo civile*, Verona, 1776–1779, 2 vols.: v. 2, "Elettricità".

Present State of Electricity,[33] terms this science as a "fashionable study" and a must-know for ladies and gentlemen.[34]

While the entertaining and social aspects of electricity should not be understated, the experiments held in aristocratic salons in the 1790s became inextricably intertwined with crucial political issues. After the beginning of the French Revolution, and in particular after 1791–92, fears of a total subversion of the social order induced local elites (including the Gazolas) to shelve the cautious reformism of the previous years. The ensuing radicalization of the scientific and political debate is particularly evident in the dispute on the so-called "animal electrometry", which raged in Northern Italy in the mid-1790s.

Controversy was spurred by the experiments on the divining rod, conducted by Pierre Thouvenel, a French mesmerist physician who moved to the Republic of Venice around 1790 with the first wave of refugees. Many of them found assistance and hospitality among the local aristocracy and were often joined by French itinerant demonstrators, who toured the Veneto conducting electrical and chemical experiments. Thouvenel believed that electricity and dowsing were both aspects of the same "universal fluid" that permeated the universe, a theory rejected by most scientific academies and by the influential Italian Society of Sciences (Società Italiana delle Scienze, also known as the "Academy of Forty")[35] because of the imponderable nature of the alleged fluid.[36] On the contrary, Thouvenel found enthusiastic supporters among the aristocracy of the many Venetian cities he visited: according to Gazola and his circle, the public experiments carried out in salons and gardens confirmed that the technocracy hostile to animal electrometry shared the same spirit of "intolerance" and "incredulity" that animated modern "materialists" and was at the root of recent upheavals.[37]

---

33 Joseph Priestley, *The History and Present State of Electricity with Original Experiments*, London, 1775.

34 *Dell'Elettricismo*: BCV, ms. 85, p. 1.

35 On the Italian Society, see: Farinella, *L'Accademia Repubblicana*..., op. cit. (n. 14).

36 Luca Ciancio, "La resistibile ascesa della rabdomanzia. Pierre Thouvenel e la Guerra di Dieci Anni", *Intersezioni* 12 (1992), pp. 267–290; Ettore Curi, "La comunità scientifica veronese e gli esperimenti di elettrometria animale alla fine del Settecento", *Atti e Memorie dell'Accademia di Agricoltura, Scienze e Lettere di Verona*, series VI, 42 (1990–1991), pp. 96–126.

37 Pierre Thouvenel, "Lettre du Docteur Thouvenel à Mr. Le Comte Gazola, Redacteur d'une brochure ayant pour titre: 'Esperienze eseguite da Pennet, in Verona'", in *La guerra di dieci anni. Raccolta polemico-fisica sull'elettrometria galvano-organica*, Verona, 1802, pp. 212–222: 215–216.

Closely tied to Thouvenel and to the French emigration, Count Giambattista Gazola placed himself at the forefront of the controversy, wrote and edited books and pamphlets in support of animal electricity, and hosted crucial experiments.[38] The family cabinet included two "large electrical machines" – the larger of which was a recent 4-*piedi* glass-plate model;[39] electrical doublers; single Leiden jars of various sizes, and batteries of six and twelve elements; six electrometers, insulating stools, capacitors, electric pistols, magic squares, electric wells and other items. Many experiments were conducted in their "Casino" – the same enclosed area on the banks of river Adige already used by Andrea Gazola in the 1740s – but normally the apparatus was placed in their main residence, in one of the five adjacent rooms dedicated to the family's collections. The other four were reserved to the paleontological museum, the library, the chemical lab and a collection of specimens from the three kingdoms of nature.

Unfortunately, I could not find any written or visual description of how the instruments were organized and displayed in the cabinet. A set of drawings of the paleontological collection (prepared around 1795 for publication in the *Veronese Ichthyology*) clearly shows that the family museum was conceived as a public space open to visitors for curiosity, discussion and courses of natural history, and regarded as an appropriate destination for the promenade of gentlewomen.[40] The museum included dozens of large marble slabs bearing impressions of fossil fishes, showcased behind glass windows which could be opened for closer inspection for the benefit of visitors or students, and paintings, representing the mountain area from which those fossils were excavated, decorated the room. Since visitors could circulate freely from one room to the next, it seems plausible that at least part of the scientific instruments were regularly on display too. The excellent condition of some of the surviving ones (such as Ramsden's quadrant, which is unlikely to have seen much practical use)[41] seems to support this inference.

---

[38] See: Gianbattista Gazola, *Esperienze eseguite da Pennet in Verona nel luglio del 1793*, Verona, 1793, and the many letters by and to Gazola published in the miscellaneous volume, *La guerra di dieci anni*…, cit. (n. 37).

[39] Gazola's electric machines no longer exist, probably because more modern ones were purchased in the 19th century for the school that came to own the collection.

[40] BCV, with no shelf mark.

[41] This quadrant saw little use even in the high school that came to own it, as it was clearly too complex for didactic purposes; the only missing parts are lenses and eyepieces, arguably recycled in simpler instruments.

The Gazolas were staunch legitimists, to the point that Louis XVI's brother (future Louis XVIII) sojourned in their "Casino" in Verona as their guest and friend for almost two years between 1794 and 1796.[42] Unsurprisingly, the family maintained a very low profile during the first years of the French occupation. Neither Bonaparte nor the Directory, however, had any intention to support genuinely democratic movements and the old urban aristocracies quickly regained their traditional pre-eminence. The "gift" by the Gazolas of their paleontological collection to the Paris Museum of Natural History in 1796,[43] and that of the experimental cabinet to the local Lyceum a few years later, should be viewed as acts of loyalty, required by the new regime in view of their complete "rehabilitation"; in the following years the Gazolas held indeed important and lucrative charges in the administration of the city. If this case study is representative of general trends, the foundation of the eight Napoleonic "*Lycées*" in Northern Italy and their equipping must have exacted a heavy toll on private scientific collections. In general, the transition from the *ancien régime* to the highly-centralized Napoleonic state looks like a turning-point for the figure of the amateur of science and the history of scientific cabinets in Italy. Gazola's fate was shared by other gentlemen natural philosophers, such as the above-mentioned astronomer Antonio Cagnoli: their collections and instruments became part of state institutes, while their owners were pressed into public service, as the old gentlemanly science got absorbed into the bureaucracy of the Napoleonic state. Eventually, few of them complained. After the initial fears, it became evident that the new *régime* intended to pursue scientific progress, while supporting the old ruling classes and enforcing social stability: in other words, the fulfilment of the program that Count Gazola and his peers had long pursued.

Appendix

In 1802 Count Giambattista Gazola offered his scientific instrument collection to the soon-to-be-founded local Lyceum,[44] which still exists as Liceo

---

[42] See: Alessandro Righi, *Il Conte di Lilla e l'emigrazione francese a Verona (1794–1796)*, Perugia, 1909, pp. 3 and 55.

[43] See: Sorbini, "Le collezioni naturalistiche...", op. cit. (n. 16), p. 97; Pietro Emilei, "Elogio Accademico del Conte Giovanbattista Gazzola", *Atti e Memorie dell'Accademia di Agricoltura, Scienze e Lettere di Verona* 14 (1834), p. 216.

[44] Tullio Ronconi, *Studi Maffeiani con una monografia sulle origini del Liceo Ginnasio S. Maffei di Verona nel primo centenario dell'Istituto*, Torino, 1909, pp. 71–72 and 101.

Classico "Scipione Maffei". Gazola's gift spelled the end of his scientific and experimental activity, but on the other hand it made it possible for at least part of his collection to come down to us. Furthermore, thanks to the documents preserved in the school archive we can outline with some accuracy the composition of the original donation and identify some of the surviving instruments.

The school started functioning in 1805 in the rooms of the former Jesuit College, but was officially inaugurated only two years later, when it moved to its present location. The first documents dealing with the scientific instrument collection date to 1807 and emphasize the need to complement it with a variety of didactic machines, since those already present were generally intended for research purposes.[45] I could not find any catalogue of Gazola's cabinet, nor of the instruments he donated to the school; the original composition of the collection can be deduced, however, from the list of the instruments it owned in 1811 and from those of the apparatus purchased by the school in the previous years, especially in 1809–1810.[46] While I cannot rule out the possibility that other local collections contributed some of the earliest instruments, there is no evidence for that and the vast majority must have come from Gazola's cabinet. The 1811 catalogue divides the cabinet of physics into the sections of "Mechanics and Dynamics", "Hydrodynamics", "Aerology", "Electricity", "Acoustics" and "Optics". It is impossible to determine whether the sections corresponded to the original arrangement, but it may not have been the case: without the recently-acquired instruments, "Mechanics" and "Acoustics" include only a handful objects that do not seem to warrant a specific section. "Optics" and "Electricity", on the other hand, were far more complete and saw very few additions; this is consistent with Count Giambattista Gazola's prevalent scientific interests. The following list includes the instruments that were certainly part of the cabinet before the new acquisitions. The object description was produced by the professor of physics, Giuseppe Zamboni, who compiled the 1811 catalogue. I have extracted and translated the items that may have come from Gazola's collection, while retaining the numbers they were given in the 1811 list.

---

45 ALM, FAS 1 – Envelope 1807–1808, 9 September 1807 (n. 17664).

46 *Elenco delle macchine esistenti nel gabinetto fisico del Regio Liceo-Convitto di Verona*, 28 February 1811: ALM, FAS 4 – 1811, sheet numbered: "1811 N° 49"; *Macchine acquistate per Gabinetto di Fisica del Regio Liceo-Convitto di Verona negli anni 1809, 1810*, undated sheet: ALM, FAS 4 – 1811. New purchases focused essentially on demonstrative apparatuses for mechanics and dynamics.

Mechanics and Dynamics

4. 2 small mills for experiments in the vacuum
5. Glass tube, for the fall of bodies in the vacuum
10. Central forces machine [i.e. centrifugal machine], with apparatus according to Nollet
12. Circular plane and double cone
21. Screw
25. Physical hammer [i.e. water hammer]

Hydrodynamics

2. N. 5 cylindrical glass jars
3. Macquer balance, extremely sensitive
4. 2 plates with communicating tubes
6. Metal jar, provided with glass vessels with tap and communicating tube
7. Nicholson and Baumé aerometers

Aerology

1. Hunter single-barreled vacuum pump, with Cavallo's improvement
2. Same as before, with two barrels
3. Armed bell jars, n. 6
4. Unarmed bell jars of different sizes, n. 12
5. Glass globe, with stop-cock
6. Two pairs of Magdeburg hemispheres
7. As before, to be placed in the vacuum
8. Circular ring for the bladder
9. Glass globe, with a bladder inside
10. Lead weights, to be placed in the vacuum
11. Heron's fountain, intermittent fountain
12. English siphon
13. N. 5 stop-cocks
14. Compression pump
16. Flintlock, to be placed in the vacuum
17. N. 6 barometers, of different kinds
19. Nooth apparatus
20. Plate with tube for measuring compressed air [i.e. manometer]
21. 1 syringe, with lead weight

Electricity

1. Two large electrical machines
2. Machine for double electricity
3. Small electrical machine
4. Electrophorus, diameter 2 *piedi*
5. Battery of 12 Leyden jars
6. 6 Leyden jars
7. Magic square
8. 2 large Leyden jars
9. Glass plate, with filings
10. Another glass plate
12. Electrometers:
    Cavallo
    Henley
    Saussure
    Bennet
    Volta
13. Electric pistol
14. Electric well
16. Aurora tube
17. Three glass jars for electricity in the vacuum [i.e. aurora bottles]
18. Thunder house
19. Three pairs of capacitors
20. Two doublers
21. Two Henley universal dischargers
22. Two insulating stools
23. Small electrical press
27. Electric chimes

Magnetism

1. Three artificial lodestones
2. Magnetic apparatus, in its box
3. Magnetometer
4. Saussure declination compass
5. Magnetic compass

Acoustics

1. Eatey machine
2. Phonometer

Optics

1. Graduate circle for reflections
2. 6 mounted lenses with coelostat
3. Mirrors (glass)
    plane-concave
    concave
    convex
    prismatic, portable

   Mirrors (metal)
    prismatic, fixed
    pyramidal
    conic
    cylindrical
    two concave, parabolic

   One concave, parabolic, made of gilded wood
4. Two polemoscopes
5. Catoptric tower
6. Camera obscura
7. Perpetual gallery
8. Glass box for refractions
9. Simple and compound microscope
10. Another compound microscope
11. Solar microscope
12. Achromatic telescope, focal length: 9 *piedi*
13. Another achromatic, focal length: 2 *piedi*
14. Another similar, made of brass
15. Astronomical quadrant, radius 1.5 *piedi* [actually, 2 English feet]
16. Transit telescope, achromatic
17. One magic lantern
18. 4 flint glass prisms
19. Two artificial eyes

COLLECTIONS OF EXPERIMENTAL NATURAL PHILOSOPHY IN EIGHTEENTH-CENTURY POLAND

Ewa Wyka

Scientific collections did not begin to emerge in Poland before the second half of the eighteenth century. Their creation was the consequence of the spreading from Western Europe of the ideas of seventeenth-century thinkers, such as Descartes, Leibniz, Bacon, Locke and Gassendi. In Poland this current of thought was called the philosophy of *recentiorum*, taken from the title of a book from 1746 called *Propositiones philosophicae ex physica recentiorum* by Antoni Wiśniewski, one of the first initiators of the Enlightenment in Poland. In the book Wiśniewski refutes scholasticism and stresses the importance of experimental physics. The inspiration for this *recentiorum* philosophy came to Poland from Christian Wolff, professor of mathematics and natural philosophy in the University in Halle, and was based on an eclectic, consensual proposal, typical of transition periods, that the truth is proven, not discovered. This philosophy tried to reconcile traditional theology with experimental science. Its reception in Poland, within the Enlightenment movement, was the first step in the replacement of scholastic methods by the gradual admission of empirical, experimental sciences into education.

Wolff's views were accepted by the Church which, as a consequence, opened the doors of Jesuit colleges to empirical sciences. The Jesuit order dominated the schools in Poland at that time, similarly to the whole of Europe. New ideas of the Enlightenment penetrated into Poland first through scientists from protestant colleges in Gdańsk and Toruń, and later through the catholic clergy educated in Italy. The foundation of the Collegium Nobilium in Warsaw in 1740 by Piarist monk Stanisław Konarski, was a landmark. It was a modern school with a well-equipped physical cabinet. The new trends did not appear in the two Polish universities of Kraków and Vilnius until 1780.[1]

[1] Władyslaw Smoleński, *Przewrót umysłowy w Polsce wieku XVIII. Studia historyczne*, Warsaw, 1979, pp. 62–104.

In order to analyse the scientific collections in Poland in the second half of the eighteenth century, we may distinguish two kinds of collections: private aristocratic collections and educational collections.

Aristocratic Collections

Collections of the Polish aristocracy were not significant. Polish aristocrats were interested in scientific innovations, but they did not create consciously-composed and carefully-considered scientific collections.[2] The most enlightened and open-minded noblemen owned only particular objects such as telescopes, theodolites and electrostatic machines. Jan Śniadecki, for instance, professor of mathematics and astronomy at Kraków University, owned a few private astronomical instruments (*Fig. 1*), among them a 12-in repeating circle by Lenoir, a reflecting circle, and an artificial horizon.[3]

The Polish aristocracy in the Age of Enlightenment aimed to increase the level of culture among artisans and to develop craft industry, rather than to start collecting scientific instruments. One of the most brilliant magnates was Michał Kazimierz Radziwiłł (1702–1762), who established factories producing glass, porcelain, furniture and textiles, as well as building a tapestry workshop. His library and collections of maps and weapons proves that he was a real connoisseur. He even edited work from his own press, was a great lover of science and art, and surrounded himself with artists and craftsmen. Michał Kazimierz Radziwiłł, and later his equally-gifted son Karol Michał (1734–1790), created art collections but did not organise physical instrument cabinets, as happened amongst aristocrats in Western Europe. At their headquarters and home in Nesvizh, there was a watchmaker's workshop, accommodating a maker called Jewna Jacobson. An arithmometer, dating from about 1770 and signed by Jacobson, has survived from this workshop.[4] It is a unique preserved instrument made by this clockmaker.

---

[2] A great number of scientific collections were created in the manor houses of German princesses, which are today located in Poland. However, these collections took on the character of renaissance *kunstkammers*. Most of the Silesian *kunstkammers* were broken up, yet some of the exhibits are known from preserved inventories. Michał Mencfeld writes about them in *Skarbce natury i sztuki. Prywatne gabinety osobliwości, kolekcje sztuki i naturaliów na Śląsku w wiekach XVII i XVIII*, Warsaw, 2010, pp. 36–69, 260–289.

[3] Antoni Szachin, Letter to the Rector of Vilnius University, 24 May 1826: Kiev, City Archive, f. 16, opis 276, dzieło 65, f. 21.

[4] Igor A. Apokin, Leonid E. Maistrow, *Razwitije wyczyslitielnych maszin*, Moscow, 1974, pp. 7–76; Leonid E. Maistrow, *Naucznyje pribory*, Moscow, 1968, p. 96.

Fig. 1. Pocket chronometer, Arnold & Son, London, 1776–1796, from a private set of instruments of Jan Śniadecki (Jagiellonian University Museum, Kraków; photo Grzegorz Zygier).

Natural history objects were collected more often than scientific instruments, as they were cheaper and more easily available. Among the Polish natural history collections, the most valuable belonged to Princess Anna Jabłonowska (1728–1800), in which thematically-arranged collections occupied four halls in Siemiatycze Palace, which contained mineralogical, zoological and numismatic exhibits, as well as a library. A rich collection of animal specimens conserved in alcohol is of particular interest.

In a letter of 29 September 1778 to Johann Bernoulli, the curator of the collection, Renata Gralath, wrote:

> The princess spares no effort or money to complete it. Last year she received a lot of boxes containing various natural curiosities from Holland [...]. Our library rather resembles a museum, and once I adorned it with over 180 jars containing tetra pods, fish, snakes, etc., preserved in alcohol, besides the birds from foreign countries, stuffed monkeys, and shells [...].[5]

[5] Quoted from: Adam Wołk, *Gabinet historii naturalnej Anny Ks. Jabłonowskiej* (online resource: http://siemiatycze.com/histor_gabinet_historii_naturalnej.html, accessed 9 January

A year later, on 30 April 1779, she wrote to J. Bernoulli:

> The Dutch have sent interesting specimens to the collection: a total of 200 different birds, which will cost 800 Dutch florins. So it is quite expensive poultry![6]

The cabinet was visited by heads of state, such as Emperor Josef II (1780), Prince Paweł, later the tsar of Russia (1782), and King Stanisław August Poniatowski (1793). After the death of the princess, the collection was purchased by Tsar Alexander I in 1801 for the cabinet of natural history at Moscow University. Together with another collection containing mostly molluscs belonging to Paul Demidow, it was destroyed during Napoleon's invasion of Russia in 1812.[7]

There were also two important collections in Gdańsk. The first belonged to Johann Philip Breynius (1680–1764), who inherited it from his father Jacob (1637–1697).[8] Breynius's exhibits occupied ten large wardrobes and contained samples of various minerals, as well as a rich collection of amber, shells and fossils. After the death of Johann Philip, the Breynius collection was bought by Catherine the Great and taken to St. Petersburg. The second was a zoological collection and, together with a botanical garden, was the property of Jacob Theodor Klein (1685–1759). He had collections of shells, mammals and selected birds, and also collected minerals and metal ores. After Klein's death, his grandson sold part of the exhibits in an auction. The library was bought by the municipal library of Gdańsk in the nineteenth century.

Among the private observatories in Poland in the eighteenth century, only one, belonging to Nathaniel Christian Wolf from Gdańsk, was of significant importance. It was situated in a building purchased by the *Societas physicae experimentalis* in Gdańsk (Gdańskie Towarzystwo Przyrodnicze) and equipped thanks to Wolf's private funds. The equipment in this observatory consisted of 17 instruments, which included a 6-ft mural quadrant by Sisson, a 4-ft meridian instrument, six telescopes (the largest

---

2013); Piotr Daszkiewicz, Aaron M. Bauer, Specimens from the Second Collection of Albertus Seba in Poland: the Natural History Cabinet of Anna Jabłonowska (1728–1800)", *Bibliotheca Herpetologica* 6, no. 2 (2006), pp. 16–20.

6 Quoted from: Wołk, op. cit. (n. 5).

7 Iwona Arabas, "Losy kolekcji księżnej Anny Jabłonowskiej na podstawie dokumentów z archiwum Rosyjskiej Akademii Nauk w St. Petersburgu", *Kwartalnik Historii Nauki i Techniki* 54, no. 2 (2009), pp. 95–108.

8 Karolina Targosz, *Jacob Breynius 1637–1697 "Botanicus Celeberrimus" w wymiarze europejskim*, Kraków, 2010; I. Arabas, *Nauka, natura i sztuka w gabinetach osobliwości*, http://www.panacea.pl/articles.php?id=275 (20 February 2012).

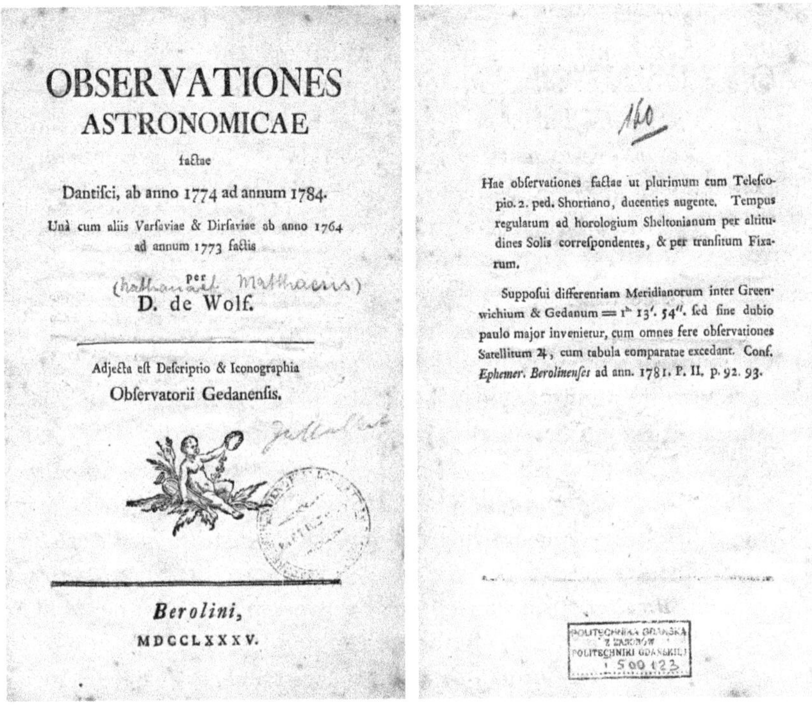

OBSERVATIONES
ASTRONOMICAE
factae
Dantisci, ab anno 1774 ad annum 1784.
Unà cum aliis Varsaviae & Dirsaviae ab anno 1764
ad annum 1773 factis
per
D. de Wolf.

Adjecta est Descriptio & Iconographia
Observatorii Gedanensis.

Berolini,
MDCCLXXXV.

Hae observationes factae ut plurimum cum Telescopio. 2. ped. Shortiano, ducenties augente. Tempus regulatum ad horologium Sheltonianum per altitudines Solis correspondentes, & per transitum Fixarum.

Supposui differentiam Meridianorum inter Greenwichium & Gedanum = 1h. 13′. 54″. sed sine dubio paulo major invenietur, cum omnes fere observationes Satellitum ♃, cum tabula comparatae excedant. Conf. *Ephemer. Berolinenses* ad ann. 1781. P. II. p. 92. 93.

Fig. 2. Title-page and description of the methods of determining the longitude of Gdańsk; Nathaniel de Wolf, *Observationes astronomicae*…, Berlin, 1785 (Biblioteka Główna Politechniki Gdańskiej, Gdańsk).

were a 4-ft telescope by Dollond and a 2-ft telescope by Short), a 1-ft quadrant by Ferguson, a Hadley octant by Dollond, a micrometer by Herschel, two regulators (by Shelton and by Magellan), a few globes by Blaeu, etc. Wolf published the results of observations carried out in Gdańsk between the years 1774 and 1784, and in Warsaw between 1764 and 1767 (*Fig. 2*). The Gdańsk observatory was destroyed during the Napoleonic war in 1813.[9]

The only aristocratic collection of instruments of real great importance was the one belonging to Stanisław August Poniatowski (1732–1798), the last Polish king (who reigned from 1764 to 1795). The beginnings of the collection go back to the period between 1765 and 1770, i.e. the years immediately

9 Nathaniel de Wolf, *Observationes astronomicae factae Dantisci, ab Anno 1774 ad annum 1784. Una cum aliis Varsaviae & Dirsaviae ab anno 1764 ad annum 1774 factis per D. de Wolf*, Berlin, 1785. In this work Wolf gives for instance the time difference between Greenwich and the observatory in Gdańsk, as well as the *elevatio poli* determined for Warsaw.

after Poniatowski's election. The collection consisted of three types of instrument: astronomical instruments, which made up the equipment of the observatory, physics instruments, kept in the royal library, and surveying instruments. This collection of more than 260 items was assembled from purchases of new instruments and the incorporation of two Jesuit cabinets, following the suppression of the Jesuit order in Poland in 1773.

The school system in Poland remained in the hands of the Society of Jesus (except for the Kraków and Vilnius academies) until 1773. After its suppression, the schools were governed by the newly-created Commission of National Education, the first ever Ministry of Education. All the material property of the Jesuits Colleges, including the physical cabinets, was given to the Commission. The two best-equipped cabinets (of Józef Rogaliński in Poznań and Stefan Łuskina in Warsaw) were incorporated into the king's cabinet.

In the royal collection there were instruments by such makers as Georg Friedrich Brander (eight drawing and measuring instruments, a vacuum pump, and a microscope), Dollond (four instruments), and Ramsden (five instruments),[10] as well as some from the workshop of Nairne & Blunt (*Fig. 3*). The relatively large number of German instruments in the royal collection reflects the German contacts of the cabinet's organiser, August Fryderyk Moszyński (1731–1786). Born in Dresden, he was the son of Fredericka Alexandrine Cosel, the natural daughter of the King of Poland and Elector of Saxony, Augustus II the Strong. He was well educated, first at military school in Dresden, and later he studied architecture under the royal architect Gaetano Chiaveri, and travelled in Europe. In 1751 Moszyński met the future king of Poland, Stanisław August Poniatowski, in Dresden and they became friends. In 1764 when Poniatowski was elected king, Moszyński was invited to come to the Warsaw court. He was given responsibility for the Royal Mint and the king's palaces and theatres, as well as for the king's collections. As an art connoisseur and science amateur, he established royal collections in both the artistic and scientific domains. He was one of a few at the royal court who knew the internal functioning of the scientific instruments market in Germany during that epoch. It can also be guessed that it was he who suggested to the King of Poland to order the scientific instruments from the workshop of Brander in Augsburg.

---

10 Celestyn Mękarski, *Catalogus Instrumentorum Mathematicorum ac Phisicorum* [sic.] *Serenissimi Stanislai Augusti Regis Poloniarum noviter per Celestinum Mękarski Perscriptus*, Warsaw, 1787: Kiev, The Manuscript Institute of the Vernadsky National Library of Ukraine, F 1, No. 5943.

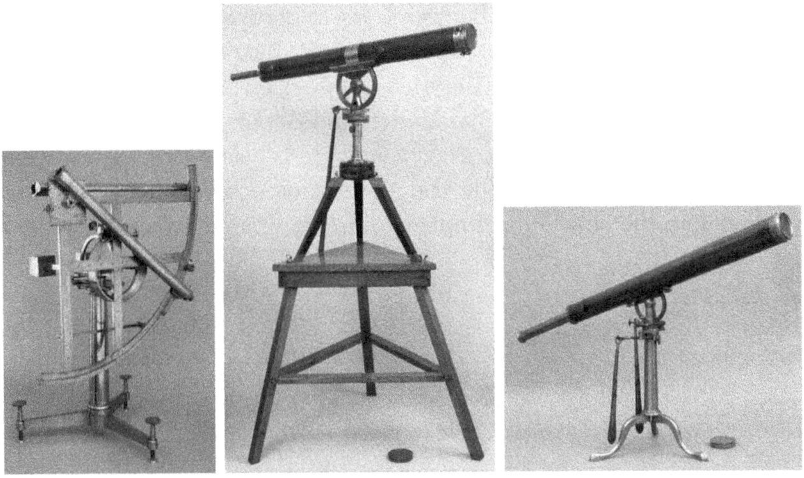

Fig. 3. Instruments from the royal cabinet of the Polish King Stanisław August, donated by him to Kraków University Observatory in the 1780s. From the left: a quadrant by J. Ramsden, a 3-ft 7-in refractor by Dollond, and a 2-ft 4-in achromatic refractor by Dollond (Jagiellonian University Museum, Kraków; photo Grzegorz Zygier).

The royal collection had a utilitarian character. Physics instruments were lent to schools as teaching aids. There was a small observatory, which functioned efficiently, where the royal astronomer, Jowin Bystrzycki, carried out typical astronomical observations for this time, as well as regular meteorological observations between 1779 and 1799.[11]

The king's library also had a utilitarian character, many of its thousands of books being devoted to science. Its catalogue contains 75 books on physics, 22 on *Ars Machinaria Practica*, 95 on mathematics and 59 on astronomy, besides many scientific books placed among volumes of philosophy. In the field of experimental physics, the king's library contained basic books by Jean-Antoine Nollet and Jan Ingen-Housz, and important handbooks of physics by Joseph-Aignan Sigaud de Lafond and John Theophilus Desaguliers, as well as works by Willem Jacob 's Gravesande and Claude-François Lambert. There were also a few works by Pieter van Musschenbroek, and others by Francis Hauksbee, Henry Power and A.G. Rozier,[12] alongside works by Cavallo, Priestley, Beccaria, Franklin, Kircher and many others.

---

11 "Dostrzeżenia Meteorologiczne czynione w Warszawie od Roku 1779 do końca Roku 1812", *Roczniki Towarzystwa Królewskiego Warszawskiego Przyjaciół Nauk* 8 (1812), pp. 129–175.

12 Among the many titles the king's library included: *L'art des expériences*, Paris, 1770, 3 vols., *Programme, ou idée générale d'un cours de physique expérimentale*, Paris, 1738, and

The library also acquired books related to particular instruments, such as those by Georg Friedrich Brander on the camera obscura, Henry Baker on the microscope, Johann Heinrich Lambert on the hygrometer, Jean-Hyacinthe Magellan on the barometer, and George Adams on surveying instruments and on the microscope.[13] There is also a very rare work by Nicolas Fuss on the telescope and the microscope.[14] Some works were dedicated to the king, for example by Magellan[15] and by Brander.[16]

Leçons de physique expérimentale, Paris, 1743–1764, 6 vols., by Antoine Nollet; *Nouvelles expériences et observations sur divers objets de physique*, Paris, 1785, by Jean Ingen-Housz; *Essai sur différentes espèces d'air-fixe ou de gas*, Paris, 1785, and *Précis historique et expérimental des phénomènes électriques…*, Paris, 1785, by Joseph-Aignan Sigaud de la Fond; *Course of Experimental Philosophy*, London, 1745, 2 vols., by John Theophilus Desaguliers; *Course of Experimental Philosophy*, London, 1745, 2 vols., by John Theophilus Desaguliers; *Philosophiae Newtonianae institutiones, in usus academicos*, Bassano, 1749 and Venice, 1749, and *Physices elementa mathematica, experimentis confirmata*, Geneva, 1748, 2 vols., by Willem Jacob 's Gravesande; *Bibliothèque de physique et d'Histoire naturelle*, Paris, 1758–1769, 6 vols., by Claude-François Lambert; *Elementa physicae conscripta in usus academicos…*, Venice, 1751, 2 vols., *Essai de Physique*, Leiden, 1751, 2 vols., *Compendium physicæ experimentalis*, Venice, 1769, and *Physicae Experimentales, et Geometricae, de Magnete, Tuborum Capillarium Vitreorumque Speculorum Attractione, Magnitudine*, Lyon, 1729, by Pieter van Musschenbroek; *Esperienze fisico-meccaniche sopra varj soggetti: contenenti un racconto di diversi stupendi fenomeni intorno la luce e l'elettricità producibile dallo strofinamento de' corpi…*, Venice, 1716, by Francis Hauksbee; *Experimental Philosophy, in three Books: containing New Experiments, Microsopical, Mercurial, Magnetical. With some Deductions, and Probable Hypotheses, raised from them, in Avouchment and Illustration of the now famous Atomical Hypothesis*, London, 1664, 3 vols., by Henry Power; *Dissertation sur les Aerostates des Anciens et des Modernes*, Geneva, 1784, by A.G. Rozier.

[13] By the aforementioned authors, the king's library included: *Kurze Beschreibung der neu abgeänderten und mit mehreren Zusätzen versehenen Camera Obscura… Als eine Zugabe zu der Anno 1769 herausgegebenen Beschreibung einer Camera Obscura*, Augsburg, 1775, by Georg Friedrich Brander; *The Microscope made easy*, London, 1754, by Henry Baker; *Hygrometrie oder Abhandlung von den Hygrometern*, Augsburg, 1774, by Johann Heinrich Lambert; *Description et Usages des Nouveaux Baromètres, pour Mesurer la Hauteur de Montagnes et la Profondeur des Mines*, London, 1779, by Jean-Hyacinthe Magellan; *Geometrical and Graphical Essays containing a Descriptions of the Mathematical instruments used in Geometry, Civil and Military surveying, Levelling and Perspective with many new Problems illustrative of each branch*, London, 1791, and *Micrographia Illustrata, or the Microscope Explained in Several New Inventions*, London, 1771, by George Adams.

[14] Nicolas Fuss, *Instruction détaillée pour porter les lunettes de toutes les différentes espèces au plus haut degré de perfection dont elles sont susceptibles tirée de la théorie dioptrique de Mr. Euler le père et mise a la portée de tous les ouvriers en ce genre. Avec la description d'un microscope qui peut passer pour le plus parfait dans son espèce et qui est propre à produire tous les grossissemens qu'on voudra*, St. Petersburg, 1774.

[15] Jean-Hyacinthe Magellan, *Notice des Instruments d'Astronomie de Geodesie et de Physique*, London, 1780; Magellan, *Description d'une machine nouvelle de dynamique inventée par M. G. Atwood, au moyen de laquelle on rend très aisément sensible les loix du mouvement des corps en ligne droite, et en rotation*, London, 1780.

[16] On Brander see: Inge Keil, Ewa Wyka, "Georg Friedrich Brander (1713–1783), Mechanicus in Augsburg and the King of Poland, Stanisław August Poniatowski (1732–1798)", in

The collection of instruments and the king's library existed almost until 1795, when, after the king's abdication, the astronomical instruments together with the royal library were bought by Tadeusz Czapski for Krzemieniec Lyceum in Lithuania, and transported there in 1805. In 1832 the school in Krzemieniec was closed and the equipment became the basis for the cabinets of the newly-established University of St. Vladimir in Kiev. The instruments were dispersed over the time, but the king's library exists and is preserved within the Manuscript Institute of the Vernadsky National Library of Ukraine.

Educational Collections

Among the collections from the second half of the eighteenth century that have been examined to date are those of the universities of Kraków and Vilnius, as well as physics cabinets of several other colleges. A full insight into the eighteenth-century Polish teaching collections requires further research into the inventories of the relatively well-equipped Jesuit and Piarist schools.

The scientific collections of the universities of Kraków and Vilnius were established as part of the reform of the Polish educational system, started in 1773 by the Commission of National Education. The observatory of the Vilnius University was the first to come into existence, in 1753, but its basic equipment was not completed until the 1770s.[17] The observatory's inventory between 1773 and 1774 contains about 120 instruments: 50 of them are astronomical, about 40 are for surveying and 30 are physics instruments.[18] Among the best instruments were an 8-ft mural quadrant by Ramsden (1777), as well as instruments made by Canivet in 1765, including a 6-ft sextant, a 4-ft meridian instrument and a 4-ft parallactic telescope. There was also a 2-ft Ramsden quadrant and two astronomical clocks by Shelton and one by Ellicott.

Ewa Wyka, Maciej Kluza, Anna Karolina Zawada (eds.), *East and West: The Common European Heritage. Proceedings of the XXV Scientific Instrument Symposium*, Kraków, 2006, pp. 29–33.

[17] Stasė Matulaitytė, *Senoji Vilniaus universiteto astronomijos observatorija ir jos biblioteka*, Vilnius, 2004, pp. 47–93; Libertas Klimka, "Senosis Vilniaus astronomijos observatorijos instrumentarijus", *Istorija* 48 (2001), pp. 41–47: 41–44.

[18] *Instrumenta do Obserwatorium JKMci Wileńskiego należące*, 1773–1774: Vilnius, Library of Vilnius University (VUB), DC 6, ff. 351–368; Marcin Poczobut, *Inwentarz Obserwatorium Wileńskiego*, 1798; VUB, VUB RS 2F-KC19, No. VI, ff. 104–108.

In 1832 the Russian authorities closed the university but the observatory still functioned until 1876. Afterwards its equipment was divided between institutions in Vilnius, Pulkovo, Warsaw and St. Petersburg. Among the 19 instruments preserved in Vilnius University, 12 come from the eighteenth-century equipment of this observatory.[19]

The physics cabinet of Vilnius University was established in 1775, and in its inventory between 1775 and 1802 there were 123 objects. Seventy of these were scientific instruments and there were 28 models of machines. Interestingly, there is a group of instruments in the cabinet made by local makers, as well as a group for demonstrating electricity, such as a voltaic pile, and a set to "electrify" different parts of the body. In the inventory there are no names of the local makers, but from the description of the instruments we can presume that some of them, like the stove models, a pyrometer, an electrostatic machine and models of machines, were executed locally. The pyrometer, for instance, was designed by Józef Mickiewicz, a professor of physics at Vilnius University, which suggests that it was also built there. Also, a model of a lift, designed by the Vilnius architect Wawrzyniec Gucewicz, and used during the renovation of Vilnius Cathedral, was certainly made in Vilnius.[20] In 1841 the cabinet was transferred to Kiev University and during the Second World War it was dispersed.[21]

The physics cabinet at Kraków University was established in 1780, after the educational reform that introduced natural history and experimental physics into the teaching programme. According to the first register of the cabinet (1784), there were around sixty instruments and about fifty accessories for two instruments: an electrostatic machine and a vacuum pump.[22] However, instruments for mechanics (about 23 items) constituted the biggest group in the cabinet. Before the reform of the University of Kraków, instruments were mainly purchased in Paris and Vienna. Other resources from the pre-reform period included a microscope with

---

[19] Matulaitytė, op. cit. (n. 17), pp. 174–179.

[20] There was no research related to the production of scientific instruments in Vilnius until now, therefore it is today impossible to determine their attribution.

[21] Libertas Klimka, Rasa Kivilsiene, "Gabinet fizyczny pod rządami profesora Józefa Mickiewicza", *Analecta* 12 (2003), pp. 26–67: 40–65; *Gabinet fizyczny pod opieką profesora Mickiewicza*: VUB, VUB F2, DC23.

[22] *Regestr Instrumentów Geometrycznych, Fizycznych, Mechanicznych, Astronomicznych w Gabinecie Collegii Physici Szkoły Głównej Koronney znaidujących się spisany Roku 1784 z wyrażeniem cechy, którą każdy Instrument iest naznaczony*: Kraków, Jagiellonian University Archive (JUA), Rkp. 398, ff. 1–49.

Sigaud de Lafond's (1758) signature, a "measuring circle" by Butterfield, and a semicircle by Canivet, as well as an air pump by Vast (1758). After the reform most of the instruments were ordered from English and French workshops (*Fig. 4*). Before 1786, a cylindrical electrostatic machine was bought from Nairne in London, with several demonstration accessories. Between 1785 and 1786, six instruments were bought in Paris: a transit instrument by Charité, a clock by Lapaute, a pair of globes by Fortin, and a meteorological barometer and a thermometer by Fortin.

In November 1790, during a meeting of the professors of the Collegium Physicum at the Academy of Kraków, a decision was made to organise the cabinet according to a physics handbook written by Sigaud de Lafond that had been published recently.[23] In 1792 a Smeaton-type air pump with accessories was bought from Nairne & Blunt of London, as well as an instrument for observation of magnetic declination. Other purchases in London were a teaching set for demonstrating magnetic phenomena, a hydrostatic balance and a set of models for teaching statics.

The person responsible for the organisation of this cabinet and the astronomical observatory was Jan Śniadecki, who had contacts among English makers. Śniadecki had a personal acquaintance with Herschel, to whom he paid a visit in 1787, and together they made astronomical observations. He also learned the method of preparing mirrors. Śniadecki also personally knew Ramsden and Dollond, and asked the latter for achromatic lenses for his parallactic telescope and quadrant, to replace old lenses.[24]

The equipment of the astronomical observatory erected in Kraków in 1792 consisted of 26 instruments, including a 3-ft quadrant by Canivet, a small quadrant by Ramsden, a 3-ft 8-in transit instrument by Charité, three clocks (two by Lepaute), and a 5-ft parallactic telescope by Dollond (*Fig. 5*).[25] Using them, Jan Śniadecki, a sophisticated observer, determined

---

[23] JUA, Rkp. A395, f. 79.

[24] "I received achromatic lenses sent from England by Dollond for my great quadrant and parallactic telescope [...] in which Dollond managed to perfectly match dimensions of focuses of these lenses with those given him by me" (odebrałem z Anglii szkła achromatyczne od Dollonda do mego wielkiego kwadrantu i lunety paralaktycznej [...] w których Dollond zupełnie trafił na wymiary ognisk ode mnie mu posłane); Jan Śniadecki, Letter to Marcin Poczobut-Odlanicki, 15 December 1792: in Mirosława Chamcówna, Stanisław Tync (eds.), *Korespondencja Jana Śniadeckiego*, Wrocław, 1954, 2 vols.: v. 2, p. 137.

[25] "Inwentarz Obserwatorium Astronomicznego zawieraiący w sobie opisanie wszystkich Instrumentów i Sprzętów w Obserwatorium znaydujących się zaczęty w Roku 1792": JUA, Rkp. WF. I-25.

Fig. 4. Air pump from the first physics cabinet of Kraków University; Vast, Paris, 1758 (Jagiellonian University Museum, Kraków; photo Grzegorz Zygier).

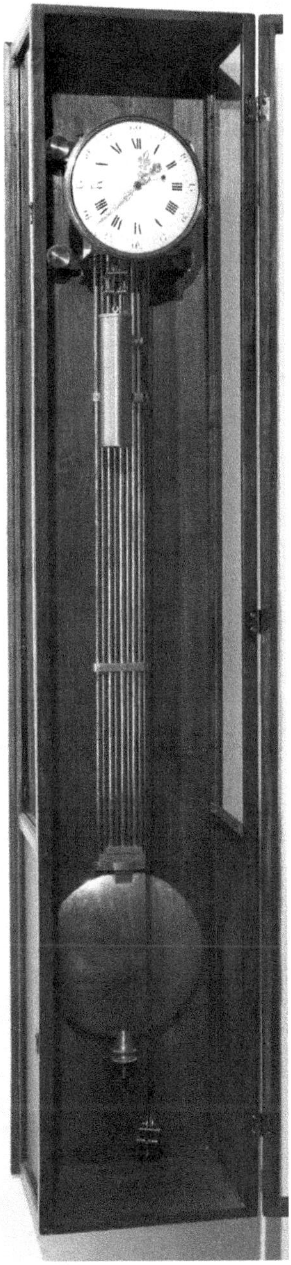

Fig. 5. Astronomical clock from Kraków University Observatory; Lepaute, Paris, 1785–1786 (Jagiellonian University Museum, Kraków; photo Grzegorz Zygier).

the coordinates of the observatory in Kraków with an accuracy comparable to modern values.[26]

The next two teaching collections were both private and institutional, because they came into existence partly due to private funding from their creators. Stefan Łuskina *s.j.* (1725–1793), Professor of Astronomy at Collegium Nobilium in Warsaw, owned the mathematical instruments that were kept in the college. A list of items has not survived, but it is known that among them were a quadrant by Canivet and "other smaller instruments".[27] In 1773, after the suppression of the Jesuit Order, he handed over his cabinet to the king's disposal and received the right to publish *Gazeta Warszawska* in exchange.

Similarly, Józef Rogaliński *s.j.* (1728–1802) established an observatory and a physics cabinet in 1762 at the Jesuit College in Poznań, partly using private funds.[28] There was also a separate hall equipped for experiments with electricity. The observatory, situated in the tower of the college building, was equipped with instruments purchased thanks to a subsidy from Maria Leszczyńska, Queen of France, and wife of Louis XV. Rogaliński's cabinet was one of the earliest modern physical collections in Poland and was modelled after French precedents. From 1759 to 1762 he stayed in Paris to study mathematics and physics, attending lectures by Jean-Antoine Nollet. The lectures and instruments used by Nollet became an example for Rogaliński's cabinet and later he would give similar lectures in Poznań. At first the lectures were for the college's students and then, from October 1766, Rogaliński gave open lectures once a week without any fee.[29]

Two Jesuits were sent by the order to help Rogaliński: Abbé Louis Sionest, Nollet's pupil and a skilled experimenter, and Jean-Baptise Delaborde, known as the inventor of an electric harpsichord.[30] Rogaliński himself never did any research in physics, but made astronomical observations.

---

[26] J. Śniadecki, "Obserwacye Astronomiczne robione w Krakowie przez Jana Śniadeckiego w Akademii Krakowskiey Matematyki wyższey i Astronomii wysłużonego Profesora", *Roczniki Towarzystwa Warszawskiego Przyjaciół Nauk* 1 (1802), pp. 462–505: 464–479.

[27] Jowin Bystrzycki, Letter to King Stanisław August, 28 January 1796: Warsaw, Archiwum Główne Akt Dawnych (AGAD), Korespondencja Stanisława Augusta, sygn. 2, ff. 15–16.

[28] Franciszek Chłapowski, *Józef Rogaliński. Uczony poznański czasów Oświecenia*, Poznań, 2007.

[29] Ibid., p. 23.

[30] Jean-Baptiste Thillais Delaborde *s.j.* (1730–1777), a French mathematician, was author of *Le Clavessin électrique avec une nouvelle théorie du méchanisme et des phénomènes de l'électricité*, Paris, 1761.

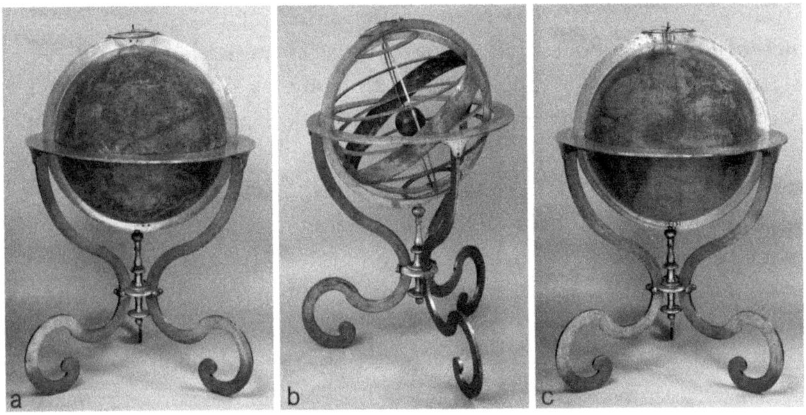

Fig. 6. Instruments from the Jesuit cabinet in Poznań: *a, c*) celestial and terrestrial globes by Robert de Vaugondy, Paris, 1751; *b*) armillary sphere by Martin aux Chevaux, Versailles, 1762. After the suppression of the Jesuit order these instruments were taken over by Kraköw University Observatory (Jagiellonian University Museum, Kraków; photo Grzegorz Zygier).

His observations of solar altitude were helpful in establishing the geographical latitude of Poznań. Based on his observation of the eclipse of the Sun on 1 April 1764, the French astronomer and Rogaliński's friend, Alexandre Guy Pingré, was able to calculate the longitude of Poznań for the first time. Other observations by Rogaliński included the Moon and the satellites of Jupiter. It is known that in the observatory there were such instruments as a 3-ft quadrant by Canivet, a 3-ft parallactic telescope by Canivet with a micrometer, a 2-ft 9-in telescope bound in black leather, two armillary spheres, a Gregorian telescope, terrestrial and celestial globes by Robert de Vaugondy (*Fig. 6*), an astronomical clock by Lepaute, and a "compteur validé".[31]

At the time of its creation in 1762, this was the best equipped observatory in Poland and, because of Rogaliński's contacts, the observatory was influenced by French science. The aforementioned French Jesuits were sent there, when the order in France began to be prohibited from doing science. Rogaliński's main work, *Doświadczenia skutków rzeczy pod zmysły podpadających*, was also influenced by the French and the frontispiece of this book is similar to that of *Leçons de physique expérimentale* by Nollet.

---

[31] Bystrzycki, op. cit. (n. 27), ff. 15–16; "Inwentarz Obserwatorium Astronomicznego ...", cit. (n. 25), WF I-25, and S.I. 443.

There was no development of original experimental physics in Poland at that time and Polish handbook authors based their work on French or English books. The inventory of the physics cabinet has not been preserved and the main source of information about its equipment is Rogaliński's physics textbook, where he describes numerous experiments carried out using these instruments.[32] Rogaliński's work consists of four volumes, making a total of 2,200 pages. Volume I (1765) concerns structure of matter, volume II (1767) mechanics, volume III (1770) practical mechanics and volume IV (1776) is generally dedicated to the art of war. The subject of the last volume was justified by the country's situation, as Poland at the time was being torn apart by war. In the same volume its author presented results of his astronomical observations. The book was written in the form of lectures and illustrated by experiments and examples of their practical use. The author was focused on practice, with detailed descriptions of various phenomena and methods of experimentation. Electricity and magnetism are not covered. Rogaliński decided to present physical phenomena without mathematical explanation and introduced many new scientific terms into Polish, although only a few of them were generally used later on. What was the importance of his textbook? It was used for teaching in the schools of Greater Poland but, although the textbook was present in the Royal Library and the Library of the Academy of Kraków, it was never used in physics education at Polish universities in Vilnius or Kraków.

It is difficult to evaluate the importance of Rogaliński's public lectures for the local community and craftsmen, as they lasted only two years (1766–1768); from 1768 Poznań was occupied by Polish insurgents and then by Russian and Prussian armies, each of them pillaging the city and using the college as their barracks. Jan Śniadecki, the future professor of astronomy at the Academy of Kraków, attended these lectures and they shaped his interest in science. The instruments from the Jesuit College in Poznań were also used for student presentations, which became a way of evaluating their knowledge at the end of the school year. It combined discussing various inventions and technical devices with experiments carried out by students using instruments from the cabinet. After 1773, the collections of both the cabinet and observatory in Poznań were distributed partly to the king's cabinet, and partly to the university in Kraków.

---

32 Józef Rogaliński, *Doświadczenia skutków rzeczy pod zmysły podpadających*, Poznań, 1765–1776, 4 vols.

The inventories of cabinets from other colleges have not yet been examined. There were about sixty of them at that time and they formed the basis of the Polish educational system in the eighteenth century. Information about their equipment can be found in the inventories of some of their physics cabinets. For example, the Lubarska School in Wołyń (now in Ukraine) ran by the Basilians around 1805, contained forty instruments, thirty of which were physics instruments and ten were instruments for surveying.[33] A small number of instruments have been preserved from the cabinet of the Jesuit College in Nysa. The college was established by Karol Habsburg, Bishop of Wrocław, in 1622 and continued to operate until 1810. The mathematician and astronomer Christoph Scheiner was its first rector. Only a dozen or so instruments from the school's cabinet have survived, including a Gregorian telescope, a nocturnal, a clinometer, a ring dial and a graphometer.[34]

A few scientific instruments have been preserved from the collection of Leopold Szersznik (1747–1814), prefect of the Jesuit College in Cieszyn, bibliophile and founder of the museum in Cieszyn.[35] Szersznik was a Jesuit, educated in the humanities, mainly in history, during his studies in Jesuit schools in Bohemia (Ołomouc and Prague). His main interests and publications concerned the history of the Slavonic peoples, but he was also interested in other subjects: between 1777–1778 he measured precipitation in the Cieszyn area and wrote a minor work about weights and measures in the city. Above all else, however, he was the author of a biographical dictionary with 108 biographies of local people of culture and science. After the suppression of the Jesuits he worked at the Clementinum in Prague and in 1775 he returned to Cieszyn, where he committed himself to social work for the city and its education. He was prefect and teacher at the local gymnasium, where he introduced modern methods of teaching, foreign languages and natural science, mainly mineralogy.[36] In 1802 Szersznik established a museum and a foundation to maintain it.

33 "Spis książek znajdujących się w Bibliotece Szkoły Powiatowej Lubarskiey", 31 February 1831: Kiev, The Manuscript Institute of the Vernadsky National Library of Ukraine, FI. 4717, ff. 19–22.

34 The instruments, which are mostly from the eighteenth century, are preserved in the collections of Nysa Museum in Nysa, Poland.

35 Irena Frencz, Maksymilian Kapalski, *Instrumenty astronomiczne i naukowe z kolekcji Leopolda Jana Szersznika w zbiorach Muzeum Śląska Cieszyńskiego*, Cieszyn, 2011.

36 Hanna Łaskarzewska, Anežka Badurová (eds.), *Ks. Leopold Jan Szersznik znany i nieznany. Materiały z konferencji naukowej pod redakcją H. Łaskarzewskiej i A. BaduroVej*, Cieszyn, 1998, p. 166.

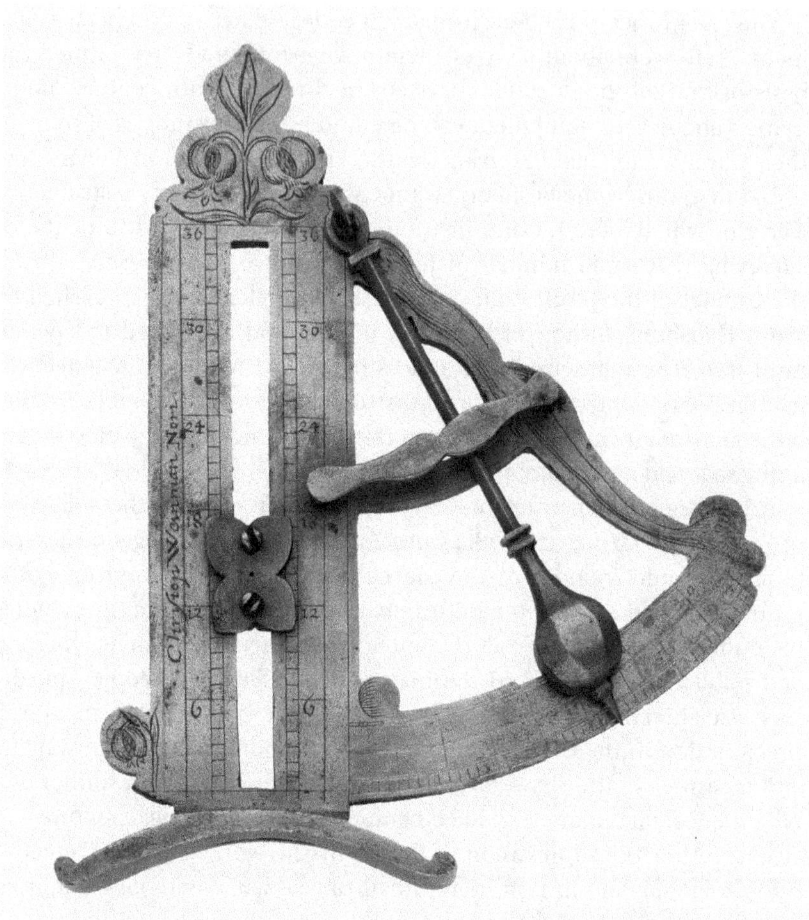

Fig. 7. Artillery level, Christoph Weinmann, Nürnberg, 1678–1688 (Muzeum w Cieszynie, Cieszyn; photo Krystian Firla).

He gathered a collection of natural history, as well as scientific instruments and models of machines, beginning his collection, mainly mineralogical, during his stay in Bohemia. Before 1801 he acquired some scientific instruments, spending 100 florins on their purchase. There were 90 instruments and models in 1805. The first inventory list created after Szersznik's death has 9,752 objects related to natural science.

No more than 30 instruments have survived from his collection today (*Fig. 7*). Among the most precious instruments are a sixteenth-century planispheric astrolabe by Adrian Zeelst (Louvain), a celestial globe by Willem Janszoon Blaeu, eighteenth-century telescopes and microscopes, and a collection of maps and books. Numerous cardboard models for the teaching of astronomy and surveying made by Szersznik and his pupils have also been preserved. These cardboard models were made for gymnasium students. Some of them have survived such as a jovilabium, a lunarium, 36 hand-painted paper models of scales with different divisions, and a quadrant made of wood and paper. They were constructed according to relief prints from Nicolas Bion's *Traité de la construction et des principaux usages des instruments de mathématique*. Further research is needed to establish Szersznik's knowledge about the production of instruments and his contacts with instrument makers. It is certain, however, that he knew the instruments offered by Brander, as evidenced by a handwritten catalogue of 43 instruments made by him. Szersznik bequeathed his library and collection to the city in 1802, beginning the history of the Muzeum Śląska Cieszyńskiego (Museum of Cieszyn Silesia).

With regard to the origin of instruments in the Polish collections, English and French work was dominant among instruments imported from Europe. Polish scientific instruments practically did not exist at that time. Single items were made in Vilnius, Kraków (by Tomasz Taborski and the Italian Gelpi brothers, for example), and Warsaw (by royal makers), but more important instruments for research and observation, and instruments for surveying, as well as more complicated physics devices like vacuum pumps, microscopes and electrostatic machines were ordered from French and English workshops (*Pl. XII*).

In eighteenth-century Poland such instruments were constructed by clockmakers. They were usually simple constructions of already-known designs, usually based on other instruments. More complex optical and measuring instruments or optical parts were usually imported, but simple tools and technical models were made by local engineers. In 1786 Marcin (?) Taborski constructed an electrostatic machine, using a glass sphere purchased in England and a set of Leyden bottles.

The maker employed at the Jagiellonian University between 1817 and 1827 was Tomasz Taborski. In his workshop a few sundials, compasses, a pendulum and a second clock (1820) were manufactured for the astronomical observatory in Kraków. He also made a sundial for the King of Poland. Taborski did not design his instruments and was a clockmaker above all else. A few clocks made by him have survived.

Those instruments made in Poland, such as the telescopes made by royal makers, were based on optics brought from Dollond's workshop. Although Poland had some good clockmakers, astronomical clocks for all Polish observatories were purchased from Lepaute, Ellicott or Shelton. The instruments in the royal collections identified as being Polish (a total of 37 pieces) are mostly composed of models of machines made by the royal maker Gotlieb Neugebauer. Polish scholars were well informed about the quality of the workshops in Europe. They visited foreign makers and corresponded with them about instruments. A rich correspondence with English instrument makers has been preserved in the Polish archives. Instruments were commissioned from Ramsden, Dollond and Nairne, as well as from prestigious French makers such as Canivet, Lepaute and Fortin.

Not many instruments from eighteenth-century Polish collections have survived. Most of the collections belonging to schools run by orders have been dispersed, although a few instruments from university collections have been preserved. In the Jagiellonian University Museum in Kraków there are about 15 instruments, while in the collections of the Vilnius University 12 eighteenth-century astronomical instruments have survived.

Summary

Despite the fact that Poland was situated on the fringes of a scientifically developing Europe, some Polish scholars made efforts to promote and practise modern science. Their activity was concentrated within structures of schools run by orders or universities and there were practically no independent scientists who had private scientific collections.

Lectures were given in major cities such as Kraków, Vilnius, Poznań and Warsaw. Physical cabinets were based in universities, using equipment stored there. It was a different situation in Warsaw, where presentations were made in public by local college professors. Warsaw was a special case, as there were public demonstrations carried out by demonstrators from abroad, who would usually offer some instruments for sale. In 1798 de Gabriel (or Degabriel), while visiting Warsaw, "gave a little course in physics". A year later (18 April 1799) he also demonstrated "some new arts highly perfected by him".[37]

Balloon flight demonstrations, which attracted many viewers, were organised in Warsaw from 1784 by Stanisław Samuel Okraszewski, although the spectacular flights with a crew aboard were later organised by Frenchman Jean-Pierre Blanchard.[38]

[37] *Dodatek do Gazety Warszawskiej* 30 (12 April 1799), p. 16. De Gabriel is known also as an instrument maker.

[38] In Warsaw they were paid performances. Blanchard came to the Polish capital together with his wife in 1790. His show attracted around 500 people. The flight was

Although the Polish aristocracy was interested in science and took part in public demonstrations, this interest did not translate into creating private scientific collections. The existing aristocratic collections were mainly of natural history and served as material for research and scientific studies.

The current research on scientific collections in the Age of Enlightenment is focussed – not only in Poland – on instruments in particular cabinets, their quality and instrument makers, as well as on the history and development of these cabinets. Other important facets have not yet been researched so thoroughly, although they are equally important, especially those concerning the development of the history of science. A first facet concerns the motives for creating such collections. In the sixteenth century, collecting various objects enhanced prestige and social standing for the aristocrats. Were not scientific collections another form of this tradition, with only a slightly different purpose? Did collecting scientific instruments stem from an inner need resulting from past traditions? Or were aristocrats really motivated by a sincere interest in science? Or was it only a matter of fashion? To get some answers, it will be necessary to analyse how these collections were eventually used.

Assuming that the creation of these collections were based on fascination with discovery in the field of physics, there is another facet worthy of analysis: to what degree was the research and demonstration potential of these collections used to promote experimental physics in the eighteenth century? It would be advisable to examine the contacts which cabinet owners had with scholars and demonstrators, to consider how these cabinets were accessible to the public, and how the instruments were used for presentations. Research in this direction would shed more light on the role which these rich collections played in the reception of modern science in the eighteenth century.

The reception process of eighteenth-century science was different in the developed countries of Western Europe, where science was born, from countries like Poland, which passively received the new ideas. This was reflected in the size and quality of the aristocratic collections of scientific instruments in different places. Comparative research would allow a wider view of how Enlightenment science spread across Europe in the second half of the eighteenth century.

---

observed from the royal observatory in the castle, and it was calculated that the balloon rose to a height of 3,975 yards.

"MONUMENTS OF SCIENCE": HOW THE TEYLER MUSEUM'S INSTRUMENT COLLECTION BECAME HISTORICAL

Martin Weiss*

Introduction

When in 1955 Jacob Kistemaker was taken on as new curator of the Teyler Foundation's physics laboratory and instrument collection in the Dutch city of Haarlem, one of his first tasks was to dismantle a Nobel prize laureate's former laboratory. From 1909 until 1928 Kistemaker's position had been filled by Hendrik Antoon Lorentz, already an internationally acclaimed Nobel prize winner when he arrived in Haarlem. No costs had been spared in providing Lorentz with the research facilities he desired.

That Lorentz accepted this position in Haarlem, resigning from a full professorship in Leiden, goes to show just how seriously the Foundation took its self-proclaimed task of facilitating experimental research, and also just how prestigious this curatorship in Haarlem was. Kistemaker in fact became the latest in a long line of eminent Dutch physicists whose research had been facilitated by the Teyler Foundation, starting with Martinus van Marum in 1784.

Kistemaker however was also the first of the curators who was not provided with his own laboratory. In this sense, his curatorship therefore marked the end of an era. What remained though was a first-rate collection of historical scientific instruments. In fact most of the instruments used by all of Kistemaker's predecessors had been preserved. Kistemaker sold off most of those he found in the laboratory, but still left a huge amount of instruments from before Lorentz's times. More importantly, these instruments were on display in the adjacent Teyler Museum. This too was funded by the Teyler Foundation. Alongside the instrument collection, this Museum housed a collection of fossils and minerals, a collection of fine art, a numismatic collection and a library.

* I would like to thank the participants of the XXX Scientific Instrument Symposium of the SIC in Kassel and Frans van Lunteren for many helpful comments on the subject of this paper.

The history of the Museum's instrument collection has been addressed previously.[1] But so far, the question of when and why the instruments were historicised has only been discussed perfunctorily. These are the issues on which this paper seeks to throw some light.

Teyler, van Marum, the Oval Room and a Laboratory

In order to understand how the Teyler Museum's collection of scientific instruments was treated and perceived at various stages throughout the nineteenth and twentieth centuries, it is important to know a little about this Museum's origins at the end of the eighteenth century.

The Museum's roots lie in Pieter Teyler van der Hulst's last will and testament. A Mennonite textile manufacturer and financier, who died a childless widower in 1778, Teyler had penned his will in 1756, stipulating that his considerable fortune was to serve charitable causes, as well as help stimulate the study of theology, and the arts and sciences.[2] He left specific instructions as to how this was to be achieved: first of all, a foundation was to be set up in his name, the Teyler Foundation, in order to look after all financial matters. This was to be run by five trustees, the first five of whom Teyler appointed posthumously through his will. The Foundation in turn was then to ensure continued funding for an alms house Teyler owned, as well as to set up two learned societies: one for theology, and another for the arts and sciences. These soon came to be known as Teyler's Theological Society and Teyler's Second Society, respectively. Teyler had stipulated that each society was to consist of six members, who should meet regularly to discuss current issues and organise annual prize essay competitions.

Finally, Teyler stated that one of his town houses was never to be sold, but was to provide a place for the Foundation's and the societies' meetings. A caretaker, preferably "a painter or other connoisseur of the arts

---

[1] See for example: Gerard l'Estrange Turner, "Teyler's Museum, Haarlem, During the Nineteenth Century", in Peter de Clercq (ed.), *Nineteenth-Century Scientific Instruments and Their Makers*, Leiden-Amsterdam, 1985, pp. 227–240; Trevor Levere, "Teyler's Museum", in E. Lefebvre, J.G. de Bruijn (eds.), *Martinus van Marum: Life and Work*, Leiden-Haarlem, 1969–1976, 6 vols.; v. 4, pp. 39–102; G. l'E. Turner, "Descriptive Catalogue of Van Marum's Scientific Instruments in Teyler's Museum", Ibid., pp. 127–396; G. l'E. Turner, *The Practice of Science in the 19th Century*, Haarlem, 1996.

[2] A transcription of the will has been published in Bert Sliggers (ed.), *Een Erfenis uit de Verlichting*, Haarlem, 2006, pp. 192–206.

and sciences",[3] was to be allowed to live in the house rent-free as well as given a handsome salary, provided he fulfilled two tasks: firstly, he was to cater for the trustees and members of the societies during their meetings and, secondly, he was to look after Teyler's private collection, which included coins, medals, books, paintings, prints, drawings and natural history specimens, and which was to be stored in the same house. This collection was to be placed at the disposal of the members of the two learned societies "at all times",[4] so that they could use it for their own research. In other words one section of his house was to serve as a repository to the learned societies.

Teyler did not however mention with as much as a word a new building or a museum – this is where another protagonist enters the stage of history: Martinus van Marum.[5] This young, ambitious, and highly talented doctor had settled in Haarlem some years before Teyler passed away, and become friends with one of the five future trustees, Jacobus Barnaart. After Teyler's death, they appear to have persuaded the other trustees not only to follow Teyler's instructions to the letter, but also to acquire new and more extensive collections, as well as to construct a new building to house these new acquisitions. They were essentially arguing that the repository Teyler had in mind should be expanded into what was first referred to as a "*musaeum*" in 1779 (the spelling "museum" was soon used interchangeably).[6] The end result was a magnificent, two-storey edifice erected behind and accessible through Teyler's old town house, which was later referred to as the "Oval Room", because of the shape of its central hall (*Pl. I*). Building work was completed in 1784 and it forms the heart of the Teyler Museum to this day.

The year the Oval Room was completed the trustees placed van Marum in charge of the Foundation's scientific collections – Teyler's natural history specimens had been sold off due to their bad condition but van Marum had already acquired a substantial amount of fossils, minerals and instruments even before 1784 – and gave him the title "Director of Teyler's Museum", thereby also cementing the new institution's name.[7]

---

3 Ibid., p. 199.

4 Ibid., p. 201.

5 On van Marum see for instance Wijnand Mijnhardt, *Tot Heil van 't Menschdom: Culturele genootschappen in Nederland 1750–1815*, Amsterdam, 1987, pp. 295–369.

6 "Vergadering", 10 December 1779: Haarlem, Teylers Museum, Archives of the Teyler Foundation (ATS), inv. no. 1382.

7 "Instructie", c. 1784: ATS, inv. no. 103.

The title of "director", van Marum's earlier pivotal involvement in the establishment of the Museum and his charismatic presence all helped eclipse the fact that he was in reality more of a co-director of the Foundation's collections, alongside the caretaker living in Teyler's town house, who already, because of his required background as a painter, was entrusted with the Foundation's collection of fine art. Yet these collections were not to be underrated. In keeping with Teyler's wish that both the arts and the sciences were to be stimulated through his bequest, the trustees provided the caretaker with a budget that rivalled van Marum's, in order to acquire works of fine art. Over the years a substantial collection of prints and drawings was brought together. The most spectacular acquisition during these early years of the Foundation's history was that of a collection of drawings formerly belonging to Christina of Sweden, which included works by renowned artists such as Michelangelo, Rembrandt, or Claude Lorrain.[8]

This duality of the Foundation's collections was to have some bearing on the instrument collection's status later on in the nineteenth century. Initially, however, all eyes were on the scientific collections and the instrument collection's *pièce de résistance* in particular: the world's largest electrostatic generator that was completed and set up in the Oval Room by John Cuthbertson in 1784. As soon as it had been set up, van Marum eagerly set to work experimenting with this new device. Just as importantly, he subsequently saw to it that the results of his experiments became widely known, sparing neither time nor effort in spreading the word that he was in charge of a well-funded museum with a machine that promised to push the boundaries of knowledge through its sheer size and power. As a result, not just he himself but also the Foundation and the Museum enjoyed growing name recognition. This in turn had the effect that the Museum, or rather its instrument collection and the electrostatic generator in particular, became an integral part of the young Foundation's identity.[9]

This did not change as van Marum continued to acquire more items for the collection over the next twenty years. Whether these were demonstra-

---

[8] On the history of this collection see Carel van Tuyll van Serooskerken, *The Italian Drawings of the Fifteenth and Sixteenth Centuries in the Teyler Museum*, Haarlem, 2000, p. 13.

[9] Travel reports from the early decades of the Museum's existence suggest visitors were far more interested in the instrument collection than any of the others. Martin Weiss, "De gang naar toegankelijkheid", *De Negentiende Eeuw* 34 (2010), pp. 269–285.

tory models for public lectures, research apparatus, or even just instruments that represented the state of the art in experimental research, they all had in common that van Marum was only satisfied with instruments of the highest quality. With the trustees' backing and the extensive network that van Marum had built on journeys through England, France and Germany, he was gradually able to acquire a first-class collection of exquisite scientific instruments.

The quality of the instruments was in fact so high that at least two visitors who published reports of what they had seen considered the instruments too precious to be used. The first of these visitors was Kaspar Heinrich von Sierstorpff, a German civil servant and junior minister from Brunswick. He came to the Museum in 1803, and complained, ironically, that such a lot of money was being wasted – to his mind – on scientific instruments; one could have spent "ten times" less on them, he argued, because not only were they unlikely to be used much, because they were so precious and it would take a long time to clean them again, but also because science was progressing so rapidly that they would soon be outdated and would then become obsolete. Once they had become outdated, he writes in his travel report, "the entire collection will someday only serve as testimony to a history of physics".[10]

Just a few years later, August Hermann Niemeyer, co-director of the Franckesche Stiftungen in Halle, said much the same, although in a slightly more positive manner. He too wrote that in the Museum he had seen "Mahogany cases full of all kinds of the most valuable physical instruments, which, with increasing perfection [*Vervollkommnung*], will be able to serve as testimony to a history of physics".[11] Considering that the earliest currently known example of a museum explicitly adopting a policy of preserving instruments for their historical value is that of the Conservatoire des Arts et Métiers in Paris in 1801, von Sierstorpff's and Niemeyer's remarks are strikingly early.[12]

Besides the collection's value however there were two other factors that would have contributed to the array of instruments on display in the Oval Room making a memorable impression. Firstly, van Marum had installed

---

10 [Kaspar Heinrich Freiherr von Sierstorpff (attr.)], *Bemerkungen auf einer Reise durch die Niederlande nach Paris im eilften Jahre der grossen Republik*, 1804, 2 vols.; v. 2, p. 559.

11 August Hermann Niemeyer, *Beobachtungen auf Reisen in und außer Deutschland*, Halle, 1820–1826, 4 vols.; v. 3, p. 151. This is a report of travels undertaken in 1806.

12 Arthur MacGregor, *Curiosity and Enlightenment: Collectors and Collections from the Sixteenth to the Nineteenth Century*, New Haven-London, 2007, p. 227.

a separate laboratory in a building adjacent to the Museum. A newspaper report in the *Algemene Konst- en Letterbode* from 1791 referred to it as "a spacious and extremely well furbished *Laboratorium* adjacent to the Foundation House [Pieter Teyler's old town house], and equipped with the necessary apparatus for physical and chemical experiments and investigations".[13] Secondly, roughly around the time von Sierstorpff visited the Museum, van Marum had an immense row with the trustees and, as a result, he reduced his activities at the Museum to an absolute minimum, basically until the end of his long life, some 35 years later.[14] As a result, the instrument collection entered what can be described a sort of "sleeping beauty state" during the first decades of the nineteenth century.

So van Marum's successor, Jacob Gijsbert Samuël van Breda, found an immensely valuable, yet somewhat outdated collection of scientific instruments, when he took over as director of the Teyler Museum in 1839. Shortly after his arrival, he wrote a four-page letter to the trustees, detailing what his plans were for the Museum, which of course included the instrument collection.[15] He stated that he saw the collection as falling into three categories. The first category he defined as those instruments that could still be used for research, the second as encompassing those instruments that could not be used for research anymore because they had become outdated, and in the third category he put all the models or demonstration apparatus that had been used for public lectures. This last group of instruments van Breda suggested one could get rid of; he was obviously not interested in using the models. The first category however van Breda wanted to keep, because he himself wanted to perform research at the Museum – that was his most likely reason for taking the job in the first place. Rather remarkably, he wanted to keep the second group of instruments as well. His argument ran as follows:

> Those [instruments] of the second kind I judge that, although of little utility value, they should be kept in the collection, they are to be regarded as monuments of science, and shall always be regarded with pleasure by every scientist [*Natuurkundigen*] who knows how much they have contributed to the progress of science.[16]

---

[13] "[...] een ruim en ongemeen wel ingerigt *Laboratorium* aan het Stichtingshuis gehegt, en met den nodigen toestel tot Physische en Chemische Proefnemingen en nasporingen voorzien geworden"; *Algemeene Konst- en Letterbode* 7–182 (1791), p. 204. I am grateful to Huib Zuidervaart for drawing my attention to this article.

[14] On this conflict see Mijnhardt, op. cit. (n. 5), pp. 326–327.

[15] Jacob Gijsbert Samuël Van Breda, Letter to the trustees of the Teyler Foundation, 28 August 1839: ATS, inv. no. 23.

[16] Van Breda, op. cit. (n. 15), p. 2.

The First Art Gallery

By 1839, therefore, it had become official policy at Teyler's not to dispense with old instruments, because of their historical value. That same year, the Museum's character as well as its outward appearance changed significantly in another way: an annex to the Oval Room was completed, the so-called "First Art Gallery" (*Eerste Schilderijenzaal, Fig. 1*). Its sole purpose was to display the collection of paintings by contemporary Dutch artists that the Foundation had successively acquired since the early 1820s, after a new caretaker, Gerrit-Jan Michaëlis, had been appointed.

On the one hand this extension to the Museum was built for practical reasons. The paintings had previously been displayed in a small and badly-lit room adjacent to the original Museum building and, as the collection grew, its accommodation would increasingly have been considered inadequate. On the other hand, however, this annex's construction and its design reflected broader developments that were taking place in the world of collections and exhibitions. More specifically, aesthetic contemplation

Fig. 1. The First Art Gallery on a drawing by Johan Conrad Greive from 1862 (Teylers Museum, Haarlem, DD042d).

of works of "fine art" had emerged as a bourgeois ideal, and exhibitions of fine art, including art museums, had taken on a role as places where visitors could not only indulge in such contemplation, but could also be seen to be doing so.[17] Put differently, over the course of the first decades of the nineteenth century, visiting an art exhibition had become a way of shoring up one's credentials as a member of the rising bourgeoisie.

A visit to this end required the visitor to follow certain rules of conduct, some of which were explicit and actively encouraged, while others were implicit and many were almost ritualistic. They included dressing properly, buying a catalogue and sporting one's knowledge of both art and other exhibitions. Such patterns of behaviour were firmly in place all over the Western hemisphere by the time the First Art Gallery at Teyler's was completed. Whilst its construction was still under way, the well-known Dutch author Nicolas Beets, a student from Leiden, who wrote under the pseudonym of Hildebrand, published an essay in the national journal *De Gids* (comparable in status to *Punch* in Great Britain), in which he mockingly caricatured imaginary visitors to the annual state exhibition of fine art in The Hague. He described one mother's disappointment at having unwittingly arrived too early: "This upsets the somewhat fashionable lady; nobody to see! nobody to see her lovely daughter!"[18]

In the same essay Beets mentions the Teyler Museum too, showing that it was clearly considered part of the established Dutch art scene: a painter who is exceedingly frustrated that the work he has contributed to the exhibition in The Hague is badly hung and not appropriately admired by both critics and visitors, complains that he had initially dreamed that "Teylers Museum will want to acquire it; the Princess of Orange will need to own it; a connoisseur will offer to invest in it with gold!"[19]

This new role of art exhibitions and museums as purveyors of social status is relevant for two reasons. Firstly on a "local" level: although the

[17] On this see for instance: Andrew McClellan, "A Brief History of the Art Museum Public", in A. McClellan (ed.), *Art and Its Publics*, Malden, 2003, pp. 1–50; Tony Bennett, *The Birth of the Museum*, New York, 1995; Carol Duncan, *Civilizing Rituals*, New York, 1995. For an overview of the Dutch situation see Annemiek Ouwerkerk, *Tussen kunst en publiek*, Zeist, 2003.

[18] "Dit valt de nog wel eenigszins wereldsche dame tegen; niemand om gezien te worden! niemand om hare lieve dochter te zien!"; Hildebrand [Nicolas Beets], "Eene Tentoonstelling van Schilderijen", in Hildebrand [N. Beets], *Camera Obscura*, Amsterdam, 1998, 2 vols.; v. 1, pp. 320–329: 325–326.

[19] "Teylers museum zal het willen aankoopen; de Prinses van Oranje zal het moeten bezitten; een liefhebber zal aanbieden het met goud te beleggen!"; Ibid., p. 322.

scarcity of source material ultimately only allows one to speculate as to what the actual intentions of both the trustees and the Gallery's architect entailed, one can clearly say that most recent standards of display were emulated in the design of the First Art Gallery, thereby creating the prerequisites for the Teyler Museum to join the ranks of this new type of institution that was beginning to emerge, i.e. the modern public art museum. The Museum was moving away from its role as an eighteenth century repository – because it housed an art collection which in turn was a result of Pieter Teyler's wish that both the arts and sciences should be stimulated.

The second reason is of a more general nature. Although aesthetic contemplation of works of art on display at an exhibition allowed visitors to affirm and enhance their social status, this process was of course reciprocal. As the imaginative artist's disappointment in Beets' essay already indicates: a work of art's cultural and monetary value could be greatly boosted by it being put on display at an exhibition; exhibitions and, by extension, museums could enhance a displayed object of art's status. Crucially, once the ritualistic codes of conduct and display strategies that allowed for this reciprocal elevation of status had been firmly established, works of art could more easily be exchanged for other objects, such as artefacts of science – for example (historical) scientific instruments – thereby enhancing the status of both the objects on display (artefacts of science) and their producers and users (scientists and engineers). This strategy was actively pursued at the beginning of the twentieth century by the initiators of the many science museums and museums of the history of science founded at this time.

International Exhibitions

This second point will resurface later on in this paper. First, however, it needs to be stressed that, despite increasing recognition of the instrument collection's historical value after 1839, in the eyes of its curators it remained first and foremost a working collection until well into the twentieth century. None of the instruments acquired until then was bought because of its historical value, but only for use in research or for demonstrating some recent advancement of science. Furthermore a substantial percentage of all instruments ever acquired was not retained in the collection. According to the inventory of the Museum's instrument collection compiled by Turner in 1976, of the 434 instruments, for instance, that van Marum listed in an inventory of his own from about 1812, as many as 181 items are lost or unidentifiable.[20] The Museum's archives contain only sporadic evidence as to when, why and by whom such instruments were disposed of. Van Breda, for instance, evidently did not hesitate to get rid of some of the instruments he himself had acquired for the collection, as Huib Zuidervaart was recently able to show.[21] In 1862 van Breda requested and received permission from the trustees to sell or dispose of "a large number of outdated physical instruments".[22] These may well have included some that his predecessor had still used; his letter to the trustees in 1839 already contained the suggestion to dispose of all of van Marum's models.

Neither did later curators hesitate to use or adapt older instruments. In 1911 Hendrik Antoon Lorentz, for instance, had a Repsold universal instrument, acquired in 1878 by one of his predecessors, Volkert Simon Maarten van der Willigen, transformed into a multi-purpose spectrometer.[23] Finally, there were other ways in which the amount of objects in the collection was reduced: in 1867 an air pump was donated to the newly-established polytechnic in Haarlem.[24]

Over the second half of the nineteenth century, however, scientific research changed profoundly. As science became ever more specialised, the equipment required to conduct experiments became ever more sensitive. Partly as a result, the distinction between amateurs and experts became more pronounced, all the more so as science professionalised and vocational training became a prerequisite to participation in any serious form of research.

These changes were reflected at Teyler's in a number of ways. First of all, whilst van Breda had still been in charge of both the geological and the instrument collection on his own, he was succeeded by two curators upon his retirement in 1864, one for geology and one for physics. The latter position was taken up by van der Willigen.

---

[20] Turner, "Descriptive Catalogue ...", cit. (n. 1).

[21] Huib Zuidervaart, "Verloren instrumenten uit het kabinet van Teylers Museum in het midden van de negentiende eeuw", *Teylers Magazijn* 96 (2007), pp. 6–11.

[22] "Wordt besloten, een groot aantal verouderde physische werktuigen, tot Teijlers Museum behoorende, welke ook door den Hr. van Breda, directeur van het Museum, voor onbruikbaar en thans onnuttig verklaard zijn, opteruimen & bij daartoe voorkomende geschikte gelegenheid te gelde te maken."; "Directienotulen", 26 September 1862: ATS, inv. no. 9, fol. 312.

[23] On the details see Marijn van Hoorn, "The Physics Laboratory of the Teyler Foundation (Haarlem) under Professor H.A. Lorentz", *Bulletin of the Scientific Instrument Society* 59 (1998), pp. 14–20.

[24] "Directienotulen ...", cit. (n. 22), fol. 392.

Secondly, the distinction between the laboratory and the Museum became more pronounced. The distinction had already been in place since van Marum's times, but by the time van der Willigen arrived, it was increasingly obvious that the Oval Room was no longer suited for most types of experimental research (although he did make use of its high ceiling for suspending a Foucault pendulum). Van der Willigen even had a new, separate laboratory building constructed in the Museum's garden.[25] With no physical connection at all to the Museum building, he could go about his spectrographical research undisturbed there. The separation of these buildings did however highlight the difference between an inaccessible research laboratory that was off-limits to anyone but experts, and the almost century-old Oval Room and the many instruments it housed, which was furthermore intimately connected to the First Art Gallery – visitors had to pass through the Oval Room to reach the Gallery. Although the laboratory and the instrument collection housed in the Museum formed an organisational unit which was distinct from the art collection, the two collections would therefore inevitably have been perceived as belonging to the same unit, which was distinct from the laboratory, namely the Teyler Museum.

Thirdly, by the fourth quarter of the nineteenth century, the increasing self-confidence of the new scientific community had brought about a widespread and growing interest in the history of science and technology, because "it helped to validate new social or intellectual pretensions".[26] The Netherlands were no exception in this respect.[27] As a result, industrial fairs such as the World Fairs increasingly included displays of historical instruments. This in turn led to renewed interest in the historical instruments from the Teyler Foundation's collection as exhibits for these historical displays. Considering that for the entire first half of the nineteenth century the Foundation had followed a strict policy of not allowing any items from any of its collections to be put on loan or even removed from its premises, it is all the more striking that the trustees agreed to provide instruments for at least two major international exhibitions.

---

25 Turner, "Teyler's Museum…", cit. (n. 1), p. 235.

26 Anthony J. Turner, "Paris, Amsterdam, London: The Collecting, Trade and Display of Early Scientific Instruments, 1830–1930", in P.R. de Clercq, *Scientific Instruments: Originals and Imitations*, Leiden, 2000, pp. 23–47: 26.

27 On the Dutch scientific community's sense of history at the end of the nineteenth century see for instance: Klaas van Berkel, *Citaten uit het boek der natuur*, Amsterdam, 1998, pp. 221–239.

The first of these was the *Special Loan Collection* held in South Kensington in 1876. Its strong emphasis on historical instruments provided the impetus for many other historical displays at later exhibitions.[28] As the catalogue stated, the organisers of the South Kensington show

> desired to obtain not only apparatus and objects from manufacturers, but also objects of historic interest from museums and private cabinets, where they are treasured as sacred relics, as well as apparatus in present use in the laboratories of the professors.[29]

The Teyler Foundation loaned them a total of ten pieces of equipment. Roughly half of these had recently been used by van der Willigen in his laboratory, the other half were sent in because of their historical value. These included a "Leyden Jar of five and a half square feet coated surface" that had been used by van Marum and had even been restored, not just repaired. The accompanying explanation in the exhibition catalogue read: "The coatings of tinfoil have been renewed recently; but all is restored in the form in which it was used by Van Marum".[30]

The Teyler Foundation's contribution to the second international exhibition, the *Paris Electrical Exhibition* of 1881, was even larger, in that the entire Cuthbertson electrostatic generator was transported to Paris and put on display there (*Fig.* 2). Already, because of its sheer size, it dominated the display of instruments from the Netherlands and its historical significance was not lost on visitors either. As much at least can be inferred from the introduction to the exhibition's catalogue, in which an explanation of the exhibition's structure contained the following passage:

> Then comes the exhibition of the Netherlands, in the midst of which has been placed the electric machine of Van Marum, almost a monument, and in any case a historical curiosity. [...] during its age, the electric machine of Van Marum was a marvel.[31]

---

[28] On the exhibition and its significance for future displays see A.J. Turner, op. cit. (n. 26), p. 27; P.R. de Clercq, "The Special Loan Collection of Scientific Apparatus", *Bulletin of the Scientific Instrument Society* 72 (2002), pp. 11–19; 73 (2002), pp. 8–13; and 74 (2002), pp. 16–23.

[29] *Catalogue of the Special Loan Collection of Scientific Apparatus at the South Kensington Museum*, London, 1877 (3rd edition), p. xiii.

[30] Ibid., no. 1367, pp. 319–320.

[31] Henri de Parville, *L'Électricité et ses Applications: Exposition de Paris*, Paris, 1882, p. 7.

Fig. 2. The electrostatic generator on display at the Paris Electrical Exhibition in 1881 (Teylers Museum, Haarlem).

A New Annex

One of the reasons the trustees agreed to part with the electrostatic generator for several months and run the risk of it being damaged on the 500 kilometre railroad journey required to transport it to Paris, was that the exhibition coincided with the construction of a major new annex to the Oval Room. This provided extra space for the instrument collection – the electrostatic generator was moved from the Oval Room to this new wing upon its completion in 1885 – as well as for the geological collections and the library. New laboratory premises were installed in a building adjacent to the new annex, replacing those originally constructed for van Marum. The new annex also included a lecture theatre that could seat about 150 people. Because space that had previously been used to store fossils and minerals was freed as a result, small semi-annual temporary exhibitions of prints and drawings from the Foundation's collection began to be held in these rooms.

Fig. 3. The entrance to the Teyler Museum after 1885 (Teylers Museum, Haarlem, AN159).

Far greater significance however lay in the new annex's design, because it changed the entire outward appearance of the Teyler Museum. Again very few sources have been preserved that could reveal why the trustees of the Teyler Foundation chose to build this annex, or why the architect chose the particular design he did.[32] But two important factors can clearly be identified as having had a profound influence on the Museum.

The first of these is that the new annex also provided the Museum with a new entrance (*Fig. 3*). Previously, access had been through Teyler's old town house; now the entrance had been moved to the other side of the premises and visitors entered through large winged doors that formed part of the new, monumental, neo-classical façade of the building's latest

[32] For the history of the annex's construction see T. van Gestel, A.W. Reinink, "Het 'Nieuwe Museum' van Teyler", in J.H. van Borssum Buisman *et alii*, *Teyler 1778–1978: Studies en bijdragen over Teylers Stichting naar aanleiding van het tweede eeuwfeest*, Haarlem, 1978, pp. 223–321.

extension. Alongside the neighbouring Dutch town houses from the seventeenth and eighteenth centuries, the new museum front stood out and could hardly be missed. Records indicate that visitor numbers increased fourfold once the new building had been opened, to more than 3,000 annually.[33]

Secondly, the new annex's architectural features and its new façade in particular, not only led to the Museum's heightened visibility, but also made it instantly recognisable as a public institution in the cultural realm. As the nineteenth century had progressed, urban landscapes had increasingly come to include monumental civic buildings such as theatres, opera houses, court houses, universities, and also museums. Entire museum quarters even saw the light of day in major Western cities. They increasingly came to be fashioned in a distinct architectural style, frequently borrowing elements from prestigious ancient – or sometimes also mediaeval – buildings.[34]

The Teyler Museum conformed to this trend. Not only its façade, but also the temple-like entrance hall with statues and reliefs allegorising various branches of the arts and sciences, as well as the well-lit exhibition area with a terrazzo floor and permanent, tall, glass showcases, in which the fossils, minerals and instruments could be displayed, were all in accordance with general, late-nineteenth-century architectural conventions of what a museum should look like. What's more, against this backdrop the guards, who now policed the Museum premises, will not have looked out of place either.

Its new outward appearance and its espousal by the general public cemented the Teyler Museum's status as a modern, public museum, moving it even further away from its initial role as a repository to Teylers learned societies than the First Art Gallery already had. Yet at the same time, as far as the curators of the scientific collections were concerned, their primary task was still to perform research. They were in fact somewhat puzzled by the large number of visitors who suddenly came into contact with their collections. The curator of geology, for instance, Tiberius Cornelis Winkler, had spent years compiling a detailed catalogue of all the fossils and minerals in the Foundation's possession, yet soon found

33 Geert-Jan Janse, "Out of Curiosity and for Instruction", in Marjan Scharloo (ed.), *Teylers Museum: A Journey in Time*, Haarlem, 2010; pp. 11–29: 23.

34 For a summary assessment of the literature on the architecture of museums housing scientific collections see Sophie Forgan, "Building the Museum: Knowledge, Conflict, and the Power of Place", *Isis* 96 (2005), pp. 572–585.

that the lay audience that now came to the Museum was at a loss when confronted with the amount of detail his catalogue provided, even though it was prominently placed on the very first showcase. In Winkler's words "it increasingly transpired that only the educated took a look at it [the catalogue], whilst the general public did not pay the least bit of attention to it".[35] He subsequently decided to distil all the information into two general guidebooks, which were first published in 1887 and were so popular that they were soon sold out.

The curator of physics, Elisa van der Ven, who had succeeded van der Willigen after the latter's unexpected death in 1878, followed suit in 1898, publishing a guidebook to the instruments on display in the Oval Room and in one of the three exhibition halls in the new annex. As van der Ven stated in his introduction to the small booklet, he had written it in "an attempt to make a visit to Teylers' unique collection of scientific instruments more worthwhile for amateurs of physics and therefore enticing".[36] But in a clear indication that van der Ven's priority had originally lain and probably still lay with expert visitors, the booklet also contained an updated inventory of the entire collection, which van der Ven had first compiled in 1882 and which included all those instruments that were not stored in the Museum building but in use at the laboratory. Van der Ven even explained that this ensured that the booklet was of use to experts and lay visitors alike.

Significantly, he also stated that lay visitors would have to pass by most of the instruments stored in the Museum, due to their lack of familiarity with the laws of physics, but then added somewhat circuitously:

> But despite this there is so much in the collection that, even if it is only because of its historical value, can elicit the interest of those too who are unfamiliar with the physical sciences, so that even just pointing them out in a better fashion did not seem like futile work to me.[37]

---

[35] "[...] bleek het hoe langer hoe meer, dat slechts geleerden er een blik in wierpen, maar dat het groote publiek er geen de minste aandacht aan schonk."; Tiberius Cornelis Winkler, "Geschiedenis van de paleontologische collectie": ATS, inv. no. 211.

[36] "[...] een poging, om een bezoek aan Teylers' eenige verzameling van natuurkundige instrumenten voor leeken op natuurkundig gebied meer vruchtbaar te maken en daardoor aantrekkelijk"; Elisa van der Ven, *Gids door de Verzameling Physische Instrumenten*, Haarlem, 1898, p. iii.

[37] "Maar niettegenstaande dat is er in de verzameling zoveel dat, al was het alleen om zijne historische waarde, de belangstelling, ook van in natuurkundige wetenschappen onbedrevenen, tot zich kan trekken, dat eene nadere aanwijzing daarvan alleen mij geen nutteloos werk toescheen"; Ibid.

This implies that history provided an angle from which van der Ven could approach the collection without losing the lay public's interest, because history formed a common denominator for all visitors – something every visitor could relate to and more easily explicable than the science behind the machines. Throughout the guidebook van der Ven accordingly repeatedly elaborated on various items' historical significance, both for research performed in Haarlem and with regard to science as a whole. The visitor's sense of history would have been augmented by van der Ven's decision to store most instruments from before 1850 in the Oval Room and all later devices in the new building. With the explanations from the guidebook, history became a little more tangible in the Oval Room.

Museum of the History of (Its Own) Science?

Van der Ven's guidebook therefore provides an important marker in the history of the Teyler Foundation's instrument collection, because it can be taken as the point at which part of the Teyler Museum became a museum of the history of science.

The crucial point is that van der Ven was explicitly catering to a lay audience. This in turn was the direct result of the fact that the instrument collection formed an integral part of the Teyler Museum as a whole, and that this museum had evolved into an almost stereotypical public museum, thereby attracting increasingly large amounts of visitors. The scientific instruments were therefore subjected to the "museum-going public" earlier than they might have been, had they not formed part of the idiosyncratic organisational structure that had sprung from Teyler's last will and testament. Van der Ven's handling of the instrument collection can therefore also serve as an example of the constitutive role of the general public at exhibitions, in the sense that his emphasis on the collection's historical dimension was a direct result of the collection's exposure to a wider, lay audience.

Yet one must not forget that the Foundation's primary focus as far as physics was concerned was still the laboratory. The trustees' aim was to facilitate research and not science classes or history courses. Their priorities became unmistakeable when they appointed Lorentz as successor to van der Ven in 1909. The laboratory premises used by van der Ven, that had been built in the 1880s during the construction of the new annex to the Museum, were given a complete and very costly overhaul. During the process they were completely sealed off from the Museum building, as a fire precaution. Until Lorentz passed away in 1928, a succession of aspiring and talented physicists were employed as assistants to the Nobel prize laureate, which effectively entailed their running the laboratory and conducting all sorts of experimental research. Neither Lorentz nor his assistants, however, actively promoted visits to the Teyler Museum as visits to a museum of the history of science. They restricted themselves to preserving, i.e. repairing and cleaning, the older instruments on display.

To draw on a comparison with a more recent class of institution, during this period the physics section of the Teyler Museum displayed many characteristics of a "visitor centre" to the Foundation's laboratory. Some of the equipment that had only recently been used was on display (van der Ven's guidebook from 1898 includes instruments acquired in 1897, for instance), scientific research was being performed nearby at a reputable laboratory, and van Marum's former "home base", the Oval Room, could be admired in its original state.

As the twentieth century progressed, however, the display of scientific instruments at the Teyler Museum gradually came to eclipse the laboratory. There were two main reasons for this. First of all, scientific research projects in general became increasingly large-scale. When the Foundation suffered a string of financial setbacks over the first decades of the twentieth century, the comparatively small laboratory could no longer compete with its larger counterparts at universities.

Secondly, the first decades of the twentieth century saw the proliferation of two new types of museum: science museums and museums of the history of science. Although all these new institutions of course had many precursors, they had a new quality to them. In order to understand the full implications of their new characteristics for the Teyler Museum, it is necessary to recall the point that was made earlier: once the ritualistic codes of conduct and display strategies that allowed for the reciprocal elevation of status of visitors and works of art at exhibitions had been firmly established, works of art could more easily be exchanged for other objects. The point is that in these new museums, products of science and technology were actively construed as cultural artefacts by constructing a museum where they could be put on display, with the explicit aim of enhancing the status of science and technology as a whole, and shoring up their credentials as academic disciplines, although this did not of course preclude other aims such as purely educational ones. As these museums proliferated, the instrument collection at the Teyler Museum would have looked less and less out of place.

One of the first and most influential institutions consciously to pursue this strategy was the Deutsches Museum in Munich. Its initiator, Oskar von Miller, had decided to become an electrical engineer upon visiting the 1881 *Paris Electrical Exhibition* – one can only speculate as to whether he might have seen the Cuthbertson electrostatic generator.[38] In 1903, after a successful career which involved installing Bavaria's first power grid, he first presented his plans for a "Deutsches Museum von Meisterwerken der Naturwissenschaft und Technik". In 1906 the cornerstone to a new building in the heart of Munich was laid. After various delays is was completed in 1925.

In 1905, when financing for von Miller's project had not yet even been fully secured, the Dutch Nobel prize laureate Jacobus van 't Hoff was invited to give a keynote lecture at the annual meeting of the new Museum's trustees. The title of his speech was *The Teyler Museum in Haarlem*, and its subtitle "The significance of historical collections for science and technology", clearly demonstrates that the Museum was perceived as a partial history museum by outsiders.[39] In a further illustration of just how much the term "museum" had come to dominate how the instrument collection in Haarlem was perceived, van 't Hoff started his speech by stating that "the expression Teyler-Museum does not correspond to the original nature of Teylers' establishment",[40] explaining how it was related to the Teyler Foundation and who Pieter Teyler van der Hulst had been.

So just a few years after van der Ven's catalogue was published, this speech and the context within which it was held provide further strong indication that the Teyler Museum was increasingly associated with the history of science. By the 1930s, when distinct museums of the history of science such as the Museum of the History of Science in Oxford, the permanent exhibition of the Istituto di Storia delle Scienze in Florence, or the Netherlands Historical Science Museum in Leiden had been founded, the Teyler Museum was one of many – with the only difference that it

---

38 Wilhelm Füßl, *Oskar von Miller 1855–1934: Eine Biographie*, Munich, 2005, p. 46 ff.

39 Jacobus van 't Hoff, *Das Teyler-Museum in Haarlem und die Bedeutung historischer Sammlungen für Naturwissenschaft und Technik*, Munich, 1908. Van 't Hoff had previously published his speech in Dutch: van 't Hoff, "Teyler's Museum en de beteekenis van geschiedkundige verzamelingen voor natuurwetenschap en industrie", *De Gids* 70 (1906), pp. 338–348.

40 "[...] der Ausdruck Teyler-Museum dem eigentlichen Charakter von Teylers Gründung nicht entspricht"; van 't Hoff, *Teyler-Museum*..., cit. (n. 39), p. 1.

inevitably represented its own history, rather than the history of science as a whole.[41]

Conclusion

The crucial factor in the gradual transformation of the Teyler Museum from a repository for two learned societies into what was essentially an art museum and a museum of the history of science, was the changing role of exhibitions and museums within society. Exhibitions of fine art in particular became places where members of the general public came to "see and be seen" and where the value of the works of art on display could be enhanced.

Because Pieter Teyler had stipulated that his fortune was to help stimulate both the arts and the sciences, the Teyler Foundation gradually acquired an extensive collection of fine art for its Museum, i.e. for the Teyler Museum, as well as providing funds and facilities for scientific research. Even though the Teyler Museum had initially become known for its exquisite instrument collection and the research performed with that at the Museum over the last decades of the eighteenth century, its growing art collection then led the Museum to become subjected to the widespread changes in the role of museums and exhibitions. This was most obviously reflected in the design of all nineteenth-century extensions to the original building of the Teyler Museum and its rising visitor numbers.

The Museum's subjection to these changes in turn allowed for a gradual reassessment of the Foundation's collection of instruments – on display at the Museum – as objects of historical value and as cultural artefacts. At the same time scientific research was still being performed in an adjacent laboratory, with the aid of new instruments acquired by the Teyler Foundation. The end result of all these developments was that the Teyler Museum became – in part – what is best described as a museum of the history of its own science.

[41] On the emergence of museums of the history of science see Jim Bennett, "European science museums and the Museum Boerhaave", in Marian Fournier (ed.), *75 jaar Museum Boerhaave*, Leiden, 2006, pp. 73–78.

THE PHYSICS CABINET OF THE ISTITUTO TECNICO TOSCANO

Paolo Brenni

This paper traces the history of the Istituto Tecnico Toscano (Tuscan Technical Institute),[1] which was founded in the mid-nineteenth century. But for a better understanding the evolution of scientific and technical education in Florence and therefore of the genesis of the Istituto, it is necessary to go back to the last quarter of the eighteenth century and briefly to trace some of the events concerning the Regio Museo di Fisica e Storia Naturale (Royal Museum of Physics and Natural History).[2]

The Regio Museo di Fisica e Storia Naturale

The Museo was founded in 1775 by the grand duke Peter Leopold of Hapsburg-Lorraine (1747–1792) and placed under the direction of the physiologist and natural philosopher Felice Fontana (1730–1805), while the naturalist and chemist Giovanni Fabbroni (1752–1822) was appointed deputy director. Peter Leopold wanted to revamp the Medici tradition and by founding the Museo he clearly showed his desire to patronize not only sciences but also the applied arts, as far as these could be fruitful for public use. Science and its applications had to be instrumental for improving the education and wellbeing of the subjects of the grand-duchy. Moreover this institution and its collection would have been a powerful vehicle of propaganda for the image of the Lorraine dynasty and for the cultural politics of the grand duke.

The Museo was located in Palazzo Torreggiani (or Torrigiani), near the residence of Palazzo Pitti (*Fig. 1*). Installed there were the collection of physical and chemical apparatus, an impressive collection of rocks and minerals, and various zoological specimens, as well as a workshop for the

---

1 Often referred to also as "Istituto Tecnico di Firenze".

2 For a detailed and a well documented history of this museum see: Simone Contardi, *La casa di Salomone a Firenze: L'imperiale e reale museo di fisica e storia naturale (1775–1801)*, Firenze, 2002; Contardi, "The museum of physics and natural history" in Filippo Camerota (ed.), *Displaying Scientific Instruments: From the Medici Wardrobe to the Museo Galileo*, Firenze, 2012, pp. 19–35. I would like to thank Simone Contardi for the fruitful discussions concerning the role of the Museo in the 19th century.

Fig. 1. The Regio Museo di Fisica e di Storia Naturale and the botanical garden in the late eighteenth century in an engraving by Aniello Lamberti (Biblioteca Nazionale Centrale, Florence, N.A. f. 6, pl. 111).

production of anatomical waxes. The Museo included also a rich botanical garden, where plants were classified following the Linnaean system. In 1780, in the tower of the Palazzo Torreggiani, the installation of an astronomical observatory (the *Specola*) was commenced. However, in spite of the acquisition of several excellent English instruments, its realisation and functioning were problematic, not only for technical reasons related to its unfavourable location and the lack of a permanently appointed astronomer, but also because of disagreements and differences between Fontana and Fabbroni.

The Museo was regularly open to the public, while access to the cabinet of physics was reserved to chosen learned people. Fontana acquired many new instruments and arranged the repair of the old ones.[3] If the simpler models and machines for illustrating Galilean and Newtonian physics were made by local artisans, who often copied the illustrations of 's Gravesande's and Nollet's treatises, the most sophisticated optical, astronomical or electrical instruments had often to be acquired abroad

[3] Between 1775 and 1807 the number of philosophical instruments preserved in the collection doubled.

(mostly in London), because nobody in Tuscany, or in Italy, could provide them. At the same time the collection of natural history and that of anatomical models were greatly increased. Fontana tried to display the collections so that they could illustrate the sciences in a coherent way, in accordance with the ideals of the Enlightenment. They had to constitute an encyclopaedia of knowledge and, thanks to their rational organisation, were expected to be an efficient learning tool for visitors.

From the 1780s onwards, the research programs (especially in electricity, chemistry and meteorology) tended to be more and more dedicated to useful discoveries. The electrical investigations were focussed on the improvement of lightning rods, while chemical research was done in order to improve understanding of respiration and the action of plants on the quality of air. For Fontana the Museo, which should have been accompanied by the institution of a scientific academy which never materialized, had to be a place of experimentation and research, where it was possible to boost the advancement of science, in order to produce useful discoveries. He did not want a purely didactic museum, essentially dedicated to teaching.

Finally for Fontana the Museo had to be a Florentine "House of Salomon". But the pursuit of the Baconian ideal encountered several difficulties. His sometimes chaotic management of the collections, his inability to control the increasing and extravagant expenditure and the completion of his several projects, his difficulties in accepting an evolving institutional structure and bureaucracy, his problematic relations with Fabbroni (who was gaining influence with the Grand Duke) and with many of his collaborators combined to marginalize Fontana. However, at the end of the eighteenth century, the Museo and its collections were among the most spectacular, prestigious and admired European institutions of this kind.

Yet the frictions and the tensions between the director and his assistant were frequent. The latter favoured a development of the physics cabinet that was more focussed on the machines, technological models and useful practical tools. Fontana, on the other hand, was more interested in demonstrational and didactic physical apparatus. Furthermore, Fontana thought it absolutely necessary to maintain the museum and its sections united in a single place and rejected Fabbroni's proposal of moving the observatory, whose location in the tower was far to be ideal. The refusal of the Grand Duke to accept Fontana's project of instituting an academy of science in the Museo marked another step in the decline of his role and influence. As Contardi clearly pointed out, in spite of the presence in the Museo of an impressive collection of apparatus, unfortunately Fontana's projects were only partially accomplished.[4] The cabinet of physics was not a great success. No important discovery was made under his direction, most of the studies were not completed or gave misleading results, and no physicist, who might have really profited from the cabinet, was trained in it. Furthermore, it had not been possible to breed a generation of instrument makers capable of working well and producing high quality instrument (this problem would persist also during the nineteenth century).

In 1798 Peter Leopold officially conferred on Fabbroni the responsibility for reordering the organization of the Museo. Formally Fontana continued to occupy his position but in practice he lost a great part of his authority. Fabbroni, as a faithful and obedient civil servant, followed the desires of the Grand Duke. From an administrative point of view, not only did he establish a better and more careful management of the personnel and of the collections, but he also introduced a stricter financial control of the expenditure. As far as museum activities were concerned, Fabbroni followed a more utilitarian perspective and tended to enrich the physics cabinet with new technological models, such as for example mills, hydraulic pumps, or textile machines, which were particularly useful in manufacturing activities, crafts and agriculture. This showed an effort to reorient the physics collection toward more utilitarian tasks, which were directly connected with the practical need of the grand duchy. Furthermore in 1807, Fabbroni advocated the presence of teachers in the Museo and succeeded in establishing a chair of anatomy and one of botany.

But the situation in Tuscany was rapidly changing. Following the Napoleonic wars, the grand duchy was dissolved in 1801 and replaced by the kingdom of Etruria, ruled by the house of the Bourbon-Parma. In 1807 the kingdom was annexed to the French empire. In the same year a new *liceo* (high school) was inaugurated in the Museo without the presence of Fabbroni, who meanwhile had become head of the Florence Mint.

The new director was Count Girolamo de Bardi, who wanted an institution dedicated to the teaching and training of young students and proposed six scientific disciplines for the high school. The collections of the Museo had to be used in stricter connection with the educational program of the different chairs. But this short experience ended in 1814, after the collapse of Napoleonic system and with the return of the Ferdinand III of Lorraine. The activity of the high school had to be suspended and the

4 See Contardi, *La casa di Salomone*, cit. (n. 2), pp. 243–244 *et passim*.

Museo, which remained as simply part of the private collections of the Grand Duke, was inactive for a period.

Fortunately it was a quite short period. In 1829 Leopold II, who had succeeded on his father's death in 1824, chose as a new director the mathematician and physicist Vincenzo Antinori (1792–1865). With him the Museo acquired a new impetus.[5] Antinori had good contacts with the intellectual elite of Florence, was involved in the activities of the most important cultural institutions of the city, and was one of the promoters of the "Congress of Italian Scientists", held periodically from 1839. Antinori not only improved and increased the collections of physics and natural history of the Museo, but also reorganised them. He believed strongly in the paramount role of sciences in education and in the development and training of human capacities. He had a public-minded vision of science and he rearranged the collections following a coherent pedagogical project. He believed in a history of science whose evolution towards a full understanding of nature was progressive and continuous. Instruments and machines belonging to the various disciplines had to be ordered following a chronological path, reflecting the phases of the evolution of physics. So the visitors to the Museo in walking through the collections, could understand the progresses of science and of its practical applications.

Together with the more modern apparatus, the Museo also preserved the ancient instruments coming from the Medici collections, as well as the relics of Galileo and of the Accademia del Cimento.[6] In Antinori's museological approach they had a fundamental role. In 1841 the *Tribuna di Galileo* was inaugurated: a monumental hall decorated with a rich iconography and displaying many instruments illustrating the most important steps of experimental science in Italy and in Tuscany in particular. In this "scientific sacrarium" as Antinori defined it, the figure of Galileo and his disciples had a preeminent role, but several other scientists from Leonardo da Vinci to Alessandro Volta were also celebrated.

Antinori not only restructured and enriched the Museo but was able also to choose some exceptional collaborators for reorganising its activities as well as its collections. In 1831 the physicist Leopoldo Nobili (1784–1835),

---

[5] On the role of Antinori in the Museo see: Simone Contardi, Mara Miniati, "Educating heart and mind: Vincenzo Antinori and the scientific culture in Nineteenth century Florence", *Archives Italiennes de Biologie: A Journal of Neurosciences* 149-4 Suppl. (2011), pp. 57–62.

[6] Most of these instruments are today preserved in the Museo Galileo. See: Camerota (ed.), *Displaying Scientific Instruments*..., cit. (n. 2).

the Italian pioneer in the researches of electromagnetism, became director of the physical cabinet and the optician, instrument maker and astronomer Giovanni Battista Amici (1786–1863) was called to be responsible of the observatory. It has to be remembered that Amici was also a microscopist and under his guidance the wax-model makers of the museum could create several models of anatomical parts and organic structures observed with the microscope.

In 1833, the lecture courses of Bardi's high school, which had been discontinued, were revived. Three chairs were instituted: physics, comparative anatomy and zoology, and mineralogy and geology; their creation was also an occasion for acquiring new instruments, models and specimens for the museum, which was becoming the hub of Italian science. In 1859 the Lorraine dynasty abandoned Tuscany, which in 1860 became part of the rising kingdom of Italy. In the same year Antinori resigned and finally the Museo and its courses were incorporated in the newly-founded "Istituto di Studi Superiori Pratici e di Perfezionamento".[7]

In 1850, at the time of the foundation of the Istituto Tecnico Toscano, the Museo was certainly an important, well-established and prestigious scientific and educational institution. However Antinori's interest in education and his pedagogical and historical approach in presenting science to the public did not really include practical applications and technology. The high school lectures held at the Museo retained the character of an academic cultural enterprise. But in the mid-nineteenth century Tuscan commerce, industry, handicraft and agriculture needed to be modernized and that could be done only with the introduction of new machines, new tools, new instruments and new manufacturing processes. Furthermore, at the same time the growing industrialisation, the institutionalisation of technical professions (surveyors, engineers, chemists, etc.), and the increasing need for specialized technicians and directors for the new mechanized manufactures, required a new kind of educational institution. The Museo and its scientific high school were certainly not sufficient and were not adapted to introducing the required technologies or to offering a modern professional training. A technical school was needed. Probably for these reasons the grand duke Leopold II decided to reorganise and merge various existing institutions for founding the Istituto Tecnico Toscano.

---

[7] This institute was a kind of "high education college" which could get the official name of "University" only in 1923.

The Istituto Tecnico di Firenze

The Istituto Tecnico Toscano, had a long genesis and a complex history and here we can trace only some of its most important steps.[8] For several decades of the nineteenth century, the Parisian Conservatoire des Arts et Métiers with its rich technical and scientific collections, which was founded in 1794, remained an admired model of an institution for high-level technical education.[9] In 1809, during the French domination of Tuscany, a Conservatorio delle Arti e Mestieri was instituted in Florence. The Conservatorio should have collected machines, models and instruments useful for industries and manufactures. It was associated with the Accademia delle Belle Arti, whose presidency had also to superintend the newly-founded institution. In fact, since its reform of 1784 the Accademia was also associated with a group of craftsmen, whose contribution could have helped to improve the state of manufactures. In spite of some resistance from the academy, which did not like the perspective of an increasing role for practitioners and craftsmen in its activities, the Conservatorio was installed in 1811 in the former cloister of Santa Caterina in San Marco Square. Together with a director, who was Giovanni Fabbroni, two professors were nominated for the two chairs of mechanics and chemistry. Furthermore, four other professors acted as consultants and about thirty

8 Unfortunately a complete and detailed history of the Istituto Tecnico has still to be written. However the most important sources of information for the history of the Istituto Tecnico and its predecessors are: *Annuario dell'I. e R. Istituto Tecnico Toscano e della I. e R. Accademia Toscana d'Arti e Manifatture Anno 1857*, Firenze, 1857; Filippo Corridi, *L'Istituto Tecnico Toscano. Opuscoli storici e scientifici*, Firenze, 1860, "Opuscolo primo: Origine e disegno dell'Istituto"; *Relazione del Ministro di agricoltura, industria e commercio sopra gli istituti tecnici le scuole di arti e mestieri le scuole di nautica le scuole delle miniere e le scuole agrarie*, Torino, 1882, pp. 51–57; Filippo Corridi, *Ricordi di fatti contemporanei*, Firenze, 1864; Paolo Galluzzi (ed.), *Scienza e Tecnica: Saggio delle collezioni scientifiche dell'Istituto tecnico Gaetano Salvemini (già Galileo)*, Firenze, 1977; Riccardo Bacci, Mauro Zampoli, *L'Istituto Tecnico di Firenze*, Firenze, 1977; Anna Gallo Martucci, *Il Conservatorio d'Arti e Mestieri, Terza classe dell'Accademia delle Belle Arti di Firenze (1811–1850)*, Firenze, 1988; Guido Gori, "The Accademia delle Belle Arti and the Istituto Tecnico Toscano 1809–1959", in Anna Giatti, Mara Miniati (eds.), *L'acustica e i suoi strumenti: Le collezioni dell'Istituto Tecnico Toscano*, Firenze, 2001, pp. 11–30; Simonetta Soldani, "Ingegneri e studi di ingegneria nella Firenze di metà Ottocento", in Franco Angotti, Giuseppe Pelosi, Simonetta Soldani (eds.), *Alle radici della moderna ingegneria: Competenze e opportunità nella Firenze dell'Ottocento*, Firenze, 2010, pp. 1–28. Other information can be found in the *Atti della Accademia Toscana di Arti e Manifatture* (1859–1864). See also: Francesco Gravina (ed.), *Le meraviglie dell'ingegno*, Firenze, 1990; Mara Miniati (ed.), *Firenze Scienza le collezioni, i luoghi e i personaggi dell'Ottocento*, Firenze, 2009.

9 For a history of the Conservatoire and its collections see: Claudine Fontanon (ed.), *1794–1994: Le Conservatoire National des Arts et Métiers au cœur de Paris*, Paris, 1994.

representatives of the most renown craftsmen were chosen as "members" of the conservatory. Finally the Conservatorio held periodical meetings and presentations, where inventors could display new or improved machines for which prizes and medals were distributed.

In 1813 the Accademia, as well as the Conservatorio, was reorganised in three classes. The first was dedicated to drawing, the second to declamation and music, and the third to the arts and crafts. With this new organisation the name of "Conservatorio" ceased to exist and was substituted by the denomination "Terza Classe dell'Accademia delle Belle Arti". In the following years, the activities of the Third Class continued and various inventions were presented but, due to the lack of money and of commitment from the government, the collection of machines was stagnating and the original target of developing technical teaching and improving the useful arts and industries seemed to be more and more difficult to reach. In 1839 the Third Class was given the task of organising (once every three years) the *Public exhibitions of the products of Arts and Manufactures*. Participation in these exhibitions (in 1839, 1841, 1844 and 1847) was not as good as expected and in the end they were not really successful.

A new momentum for technical instruction in Florence started in 1850 with two decrees of Leopold II. With them the Grand Duke of Tuscany decided to separate completely the "technical schools of arts and manufactures" from the Accademia, and to nominate Filippo Corridi (1806–1877) director of these schools (*Fig.* 2). One of the first steps of the new director was to rename the schools "Istituto Tecnico" (a denomination which became official from 1853). The energetic Corridi was the prime mover of an ambitious project and it is worth spending a few words on his figure.[10] Born in Livorno, after having studied in Pisa, he became professor of mathematics. Well aware of the problems related to popular education and the training of apprentices, he wrote several didactic treatises for elementary schools. In 1839, he was secretary of the first Congress of Italian Scientists and in 1843 became the preceptor of the elder son of the Grand Duke and superintendent of the Florentine Società per gli Asili Infantili (Society for Children's Schools). In this position he also created a society for technical education, which provided a basic training for boys who wanted to became craftsmen. More and more interested in the industrial

[10] For a biography of Corridi see S. Soldani, "Filippo Corridi", in *Dizionario biografico degli Italiani*, Rome, 1960–: v. 29, pp. 520–526 (on-line version: www.treccani.it/enciclopedia/filippo-corridi_%28Dizionario-Biografico%29/, accessed 2 January 2013).

Fig. 2. Portrait of Filippo Corridi by Francesco Boggi, 1839, from *Album di 57 ritratti di scienziati italiani*..., Pisa, 1841 (Biblioteca Nazionale Centrale, Florence, Magl. 14.2.66).

development of the transalpine countries, he directed and organised the exhibitions of natural and industrial products of Tuscany in 1850 and 1854 and, thanks to his experience, he also was the commissioner responsible for the participation of Tuscany in the universal exhibitions of London in 1851 and of Paris in 1855.

In 1850 Corridi found a very discouraging situation. The courses of the technical schools had been suspended, the academic activities were almost irrelevant, and the collections were poorly equipped. Instruments and machines had been neglected and were in poor condition, the workshops and the laboratories almost nonexistent. Corridi's ideas were clear. He had two French models in mind: the Conservatoire des Arts et Métiers of Paris and the School of La Martinière in Lyon, both of which he had had the opportunity to visit. In spite of his important involvement with international and national exhibitions, Corridi worked hard for the creation of the reformed Istituto for which he obtained the large cloister of the "*Cavalierato di Malta*". In 1854 the official regulations of the Istituto were finally ready and in 1856 the professors of geometry, physics, chemistry, mechanics and natural history were nominated.

Finally in 1857 the Istituto was officially inaugurated by the Grand Duke in the presence of the most important representatives both of the political as well as the scientific and industrial communities of Tuscany. In his opening speech Corridi clearly pointed out that the Istituto was not mainly intended for promoting "high speculation" but for introducing scientific principles into industries, manufactures and useful arts. In modern terms we can say that the Istituto had to be devoted to applied science and technology. The role it performed should have been that of an efficient polytechnic school, which, due also to the lack of a solid university tradition in Florence, represented the city's major venue for technical-scientific study and teaching, supported by a staff of illustrious professors.

The yearbook of the Istituto for 1857 clearly defined its structure,[11] which also included some of the activities and structures that had been organised (not always successfully) in the previous decades, namely the technical schools (with the chairs of descriptive geometry and technical drawing, physics and applied physics, mechanics and applied mechanics, natural history applied to useful arts, applied chemistry and metallurgy), the academy of arts and manufactures, the technological museum, the mechanical workshop, the library and the exhibitions of the natural

[11] See *Annuario…*, cit. (n. 8), pp. 39–61.

and industrial products of Tuscany. From the beginning, the courses were successful and the physics lessons of Gilberto Govi attracted an impressive audience. The French journal *Cosmos* (edited by the Abbé Moigno), in commenting on the opening of the Istituto, reported that the lecture room was too small for accommodating the crowd wanting to follow the lessons.[12] In the following years, thanks to the determined direction of Corridi, the Istituto became one of the most important and well-structured institutions of this kind. The library and especially the collections were enormously enriched. Thanks to his local and international acquaintances and because of the many good relations established during his work for the Tuscan exhibitions and for the universal ones, Corridi could not only acquire but was also able to obtain as presents (both from private as well as from public institutions) a large number of natural specimens, sample of raw materials, chemicals, manufactured products, instruments, tools, models and books. In 1859 the first graduates left the Istituto: six came from the physical-mathematical section and four from the surveying and geometry section.

But the "soft revolution" of spring 1859 suddenly marked the end of the Lorena dynasty in Tuscany and was followed by the institution of a provisional government. At the end of the year Corridi was firmly invited to resign. In spite of his remarkable achievements and in spite of the fact that various accusations that had been put forward to justify his forced resignation were sometimes pretexts and were never really proved,[13] Corridi had to quit the direction of what can be considered his creature. He had been one of the heralds of the Lorraine restoration and was too much involved with the old regime to maintain his official position with the new rulers.

The Casati decree of 1859 (which reformed the school system) and the creation of the above-mentioned Istituto di Studi Superiori, triggered a series of continuous changes in the organisation of the Istituto which in 1861 came under the responsibility of the Ministry of Agriculture, Industry and Commerce of the newly-founded kingdom of Italy. After a period of stagnation and uncertainty, and thanks also to the fact that Florence became capital of Italy between 1865 and 1870, the Istituto acquired a new momentum. With the help of strong financial support from the

---

12 "Inauguration de l'Institut Technique Toscan", *Cosmos* 10 (1857), p. 338.

13 The most ignominious accusation made to Corridi was to have imported a guillotine from France, following the short reintroduction of the death penalty in Tuscany in 1852; see Corridi, *Ricordi...*, cit. (n. 8), pp. 1–73.

Fig. 3. The façade of the Istituto Tecnico Toscano in Via del Mandorlo (today Via Giusti), Florence, ca. 1930 (Fondazione Scienza e Tecnica, Florence).

municipality and the province, the Istituto flourished once again. In 1880 responsibility passed to the Ministry of Education and in 1883 the name was changed to "Istituto Tecnico Galileo Galilei". In the same decades, several changes happened in the structure and organisation of the school. Because of its success, the Istituto needed a larger and more modern building and in 1891 it was transferred to new premises (*Fig. 3*) built in Via del Mandorlo (today Via Giusti). This building was further enlarged around 1910. Finally, in 1923 the school reform introduced by the philosopher Giovanni Gentile (1875–1944), Minister of Education in Mussolini's government, profoundly changed the characteristics of the "Galilei", which lost the physical-mathematical section and became essentially an "Istituto tecnico commerciale e per geometri".[14] This denomination was officially adopted ten years later.

[14] The Italian term "*geometra*", which can be hardly translated with precision, indicates a profession whose activities can include surveying, building planning and construction of buildings, direction of building yards as well as valuation of real estates and lands.

The Physics Cabinet

From the 1850s Corridi worked hard to enrich the several collections and to equip the scientific cabinets of the Istituto. Here, we will focus our attention on the physics collection, which in the second half of the nineteenth century certainly was one of the best of its kind in Italy. The physics cabinet needed a large number of instruments, but the Italian instrument making industry was quite poor throughout the nineteenth and the early twentieth century. The situation in fact was similar to that of the eighteenth century. For several reasons, which have been analysed in detail elsewhere,[15] a very few Italian makers, often working for the universities or the astronomical observatories, could manufacture relatively simple instruments, which were often copied from those produced abroad. In any case they were only able partially to supply a local market. One of the few good instrument makers in Tuscany was Mariano Pierucci, who had his workshop in Pisa and worked for the local university. Pierucci supplied some apparatus for the Istituto, among which a very fine Atwood's machine of French design.

Consequently, it was common practice to acquire instruments made in England, France and, from the end of the century onwards, also in Germany.[16] Corridi, was in an ideal position to "shop for instruments" abroad. He could profit from his international relations and from his role as commissioner for Tuscany at the universal exhibitions of 1851 in London and of 1855 in Paris, by meeting the most renowned makers and examining their best instruments. In the period between the 1830s and 1880s French philosophical instrument makers were dominating the continental market and it is not surprising that Corridi, and later his successors, acquired so much apparatus in Paris. Since the early 1850s, the collection was constantly enriched with excellent instruments and not with cheap didactic toys. When Corridi left the Istituto in 1860, he proudly listed some of the best apparatus in the cabinet, such as a cathetometer and a dividing engine (*Fig. 4*) made by Perraux in Paris (the latter had been expressly ordered for the Istituto), a large Natterer machine for liquefying

---

15 See for example: Paolo Brenni, "Italian Scientific Instrument Makers of the XIXth Century and their Instruments", in Peter Richard de Clercq (ed.), *Nineteenth-Century scientific instruments and their makers*, Leiden-Amsterdam, 1985, pp. 183–203; Paolo Brenni, Massimo Misiti, "Costruttori italiani di strumenti scientifici del XIX secolo", *Nuncius, Annali di Storia della Scienza* 1 (1986), pp. 141–184.

16 About the European industry of scientific instruments in the 19th century see: De Clercq (ed.), op. cit. (n. 15).

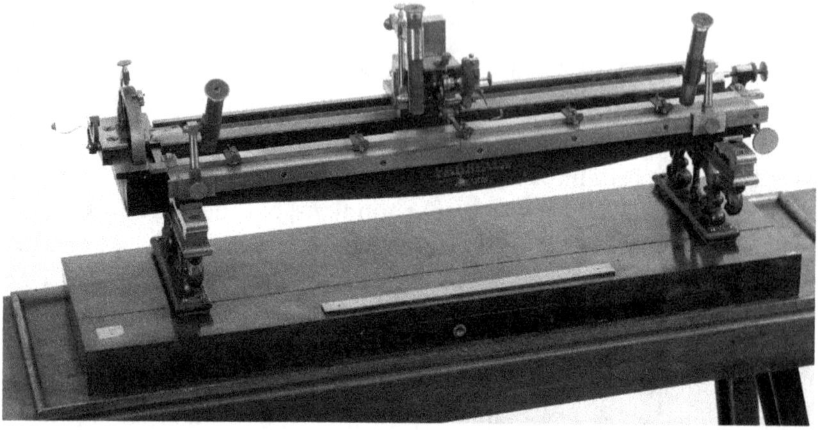

Fig. 4. The linear dividing engine made by Perreaux in Paris, ca. 1855 (Fondazione Scienza e Tecnica, Florence).

gases (Deleuil), a sophisticated optical bench, a Silbermann heliostat, a laboratory arc lamp with several accessories and a solar microscope, all made by Duboscq, a Wheatstone wave machine, a large induction coil by Ruhmkorff and an Amici microscope.[17] At the same time the electricity for the cabinet and the lecture rooms was provided by a battery of 90 Bunsen elements (50 of them had been made locally by the chemist Carrarasi of Florence, while the others had been acquired from Deleuil in Paris). Furthermore, the cabinet was also supplied with gas lighting.

If most of the sophisticated instruments were acquired abroad, some simpler apparatus was made in the workshop annexed to the collections, which in the grand project of Corridi was supposed to produce some of the apparatus for the cabinet (and for other institutions in Tuscany), as well as to do the necessary repairs. It was equipped with three lathes of different sizes and a large number of hand tools. Before 1860, in the first years of the Istituto, the workshop, which was under the responsibility of the mechanic Raffaello Turchini, was quite active. At the beginning of 1860 almost 40 instruments (and various tools) had been made for the Istituto, another 25 for various other schools and institutions, and some other 25 further instruments were not yet completed. In fact they were not really precision instruments but most of them were for teaching (mechanics, acoustics, thermology and hydraulics; *Fig. 5*). Nevertheless the good

[17] See: Corridi, *Ricordi*…, cit. (n. 8), pp. 131–132.

Fig. 5. Above: compression pump and compression fountain made in 1854 by Turchini in the workshop of the Istituto Tecnico Toscano. Below: the brass plate on the apparatus (Fondazione Scienza e Tecnica, Florence).

quality of Turchini's workmanship can be seen in some of his apparatus, which is still preserved today in the collection, today at the Fondazione Scienza e Tecnica of Florence.

However by the mid-1860s the activities of the workshop had slowed down. Around 1864 a group of people, including the instrument maker Giuseppe Poggiali and the astronomer Giovanni Battista Donati, had founded a small company in Florence for the construction of scientific instruments. After some negotiations, Donati and his associates were allowed to occupy and use the workshop of the Istituto, where they remained from 1868 to 1870, before moving into some new premises. At the same time, probably in 1869, the company officially assumed the name of "Officina Galileo".[18] It was the beginning of one of the most important Italian precision industries, which developed significantly in the twentieth century.[19] In the following decades the workshop continued to be active for the needs of the cabinet and Turchini made various special pieces of apparatus required by the professor of physics.

Over the decades the Istituto attracted several distinguished and well-known physicists, who occupied the chair of physics and were at the same time directors of the cabinet. Notable among them are Gilberto Govi (director between 1856 and 1860), Tommaso del Beccaro (1860–1865), Emilio Villari (1866–1871), Antonio Roiti (1871–1878), Antonio Bartoli (1878–1888) and Eugenio Bazzi (1888–1919).[20] After a few years spent in the Istituto, most moved to different Italian universities. However, as directors of the cabinet, their role was not limited to teaching physics and technology but they were also free to carry on their personal researches. In fact, the presence in the cabinet of several very sophisticated and expensive instruments (high-precision electrical and optical measuring apparatus, powerful electromagnets, chronographs, large calorimeters, etc.), which were certainly not necessary for teaching, was justified only if we admit that they were used for research work. Several original instruments and prototypes remain today as material witnesses of these studies; others unfortunately are lost. For example, in the collection at the Fondazione

[18] The original name of the firm, "Officina Galileo", was transformed later into "Officine Galileo".

[19] The intricate beginning of the Officina Galileo is carefully reconstructed in detail by: Alberto Meschiari, *Come nacque l'Officina Galileo di Firenze: Gli anni 1861–1870*, Firenze, 2005.

[20] For their short biographies and bibliographies see: Floriano Cajori, *Storia della fisica elementare con l'evoluzione dei laboratori fisici*, Palermo, 1930 (2nd ed.). About Eugenio Bazzi: www.carlobramantiradio.it/bazzi.htm (accessed 2 January 2013).

there is an instrument conceived by Villari for determining the dilatation of wood cut parallel or perpendicular to its fibres. Preserved from Roiti's work is a curious device for the study of induction phenomena in solenoids, as well as an instrument for visualising the effect of traction on metallic wires. From the work of Bartoli come a special voltmeter, part of an apparatus for demonstrating a theorem related to electrostatic induction, and various fragments of a complex apparatus used for the determination of the mechanical equivalent of heat from the rise in temperature of mercury flowing in a tube. From Bazzi there is a phonedoscope (of a type that he had invented),[21] some electric relays for telegraphy, apparatus for studying and recording the composition of the movements of two pendulums (*Fig. 6*), and a device for observing the shape of a liquid vein flowing from a vessel. The instruments used (or conceived) for the personal researches of the physicists who worked in the Istituto were often described or mentioned in the scientific articles they published.[22]

In the late 1850s (and probably also later) the textbook adopted for the physics courses at the Istituto was the *Traité élementaire de physique* (Paris, 1856, sixth edition, or its Italian translation) by Adolphe Ganot. This treatise was one of the most popular and most used textbooks of elementary physics of the nineteenth century.[23] First published in 1851, it reached a 25th French edition before World War One. It was also translated into about ten different languages and hundreds of thousands of copies were sold. It was used in a very large number of schools and institutes all around Europe and America, where it became a standard work for more than half a century and contributed to shaping physics teaching as well as physics cabinets.

Ganot's treatise presented a purely descriptive physics, which was almost completely deprived of mathematical formalism, while it reserved a very important place for experiments, instruments and the technical applications of physics (steam engine, telegraphy, photography, etc.). But in spite of the fact that we know the textbook was used in physics, it is much more difficult to know whether and how the hundreds of didactic instruments apparatus of the cabinet were effectively used during

---

21 The phonendoscope is a kind of stethoscope that intensifies auscultated sounds.

22 Most of them can be found in the *Annali del nuovo Cimento*, or in the *Atti della Regia Accademia dei Lincei*, which certainly were the most important Italian scientific journals at the time.

23 On Ganot's treatise see for example: Françoise Khantine-Langlois, "One Century of Physics through a Popular Textbook: Ganot's Treatise of Physics", www.societechimiquedefrance.fr/IMG/pdf/Langlois_Ganot_SFC_trad_anglais.pdf (accessed 2 January 2013).

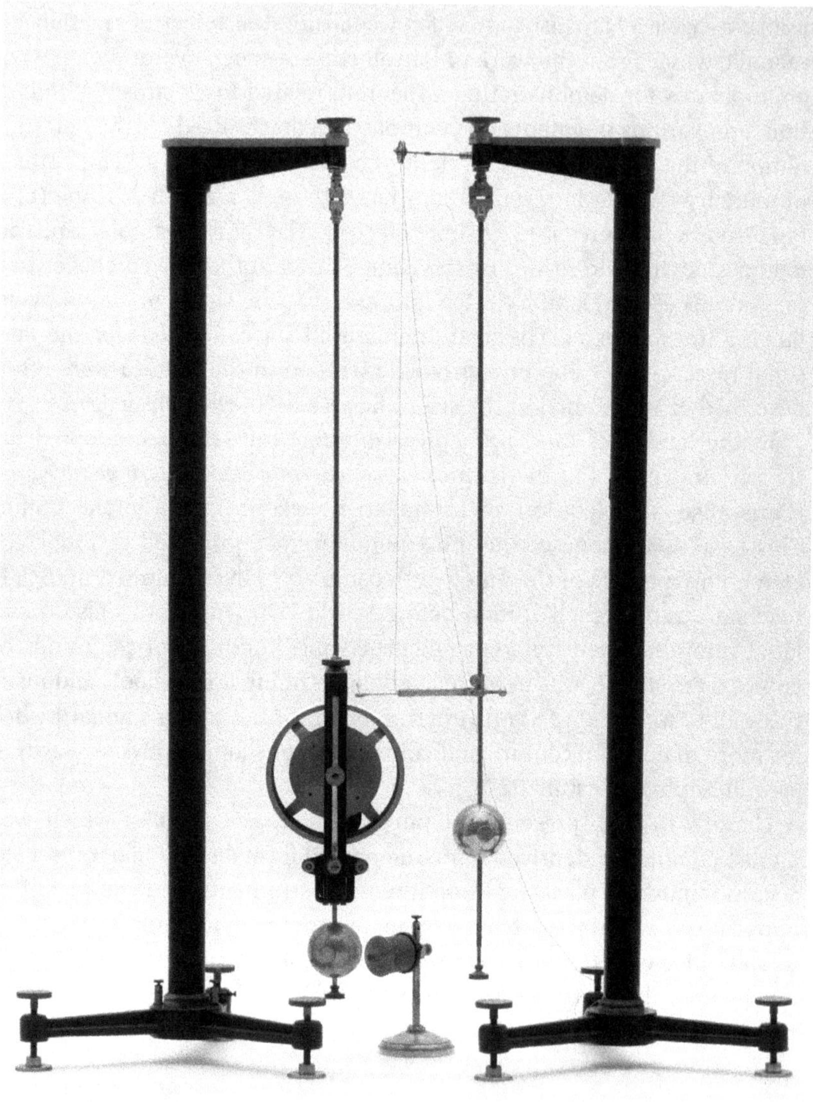

Fig. 6. Bazzi's apparatus for recording the movements of two pendulums (Fondazione Scienza e Tecnica, Florence).

the lessons. That is a major question when we study didactic collections of the past. We could answer to it only if we had notes and journals of the professors, as well as exercise and workbooks of students. Unfortunately, apart of a few handwritten sketches of experimental settings prepared for the lessons, we could not find such documents. It is therefore very difficult to understand which instruments were really regularly used and which ones were simply shown, described and explained as a three-dimensional illustration, but with no real demonstrations being performed with them. Some instruments were conceived for performing experiments that were far too time-consuming (calorimetry, for example) or too complicated for standard lessons. It is not uncommon to find instruments in such collections that had been acquired just because they were iconic: they helped to affirm the importance and the wealth of a scientific institution. Furthermore, some instruments were probably purchased because it was necessary to spend the allocated money. There is evidence that some of them were never or very rarely used.[24]

The cabinet of mechanics, which was placed under the direction of Nicola Collignon (1820–1891), was also largely developed in the 1860s and 1870s. To the old machines coming from the Conservatorio di Santa Caterina (a few of them had been originally made for the Regio Museo di Fisica e Storia Naturale), hundreds of new models and devices were added. Models of machines were common in the physics cabinet from the mid-eighteenth century. Mills, pile drivers, cranes, waterwheels, pumps, etc., were examples of the possible practical application of fundamental physical laws of mechanics and hydraulics. During the nineteenth century the number of models in physics collections was greatly increased with the invention and diffusion of new machines in an era of fast-growing industrialisation. This was particularly true for schools, which, like the Istituto, emphasised a technical approach to physics.

As for the cabinet of physics, many valuable models and apparatus of the mechanics cabinet, such as a beautiful Watt engine and a Watt indicator, a Reichenbach water column machine, dynamometers of various types, several models of turbines, waterwheels and hydrometric instruments, were acquired from Salleron in Paris. A large collection of kinematical models, made following the direction of the German Ferdinand

24 For the history and use of nineteenth-century didactic physic apparatus, see: Paolo Brenni, "The evolution of theaching instruments and their use between 1800 and 1930", in Peter Heering, Roland Wittje (eds.), *Learning by Doing Experiments and Instruments in the History of Science Teaching*, Stuttgart, 2011, pp. 281–315.

Fig. 7. The gallery of mechanics, ca. 1930 (Fondazione Scienza e Tecnica, Florence).

Redtenbacher (1809–1863), one of the fathers of science-based mechanical engineering, came from the instrument maker J. Schröder of Darmstadt University. Furthermore, other models related to applied mechanics and hydraulics (such as a series of pumps of different kinds) were made in the workshop of the Istituto. In 1879 the original chair of mechanics and hydraulics was suppressed and its collection was incorporated into the cabinet of physics (*Fig. 7*).

Great efforts in providing the best instruments for the Istituto were constantly carried on by the physicists who were successive directors of the cabinet. The increasing importance of the German precision industry in the last decade of the nineteenth century and in the first years of the twentieth was clearly reflected in the collection. With the spectacular industrial development of a united Germany after 1870, the instrument makers of Berlin, Frankfurt, Chemnitz, Köln, München, Dresden, etc., began to flood the market with their excellent products, while Parisian makers were loosing part of their importance. In fact, from the 1890s the acquisitions of apparatus made by Max Kohl, Leybold Nachfolger, Hartmann & Braun or Franz Schmidt & Haensch, just to mention the most prominent ones, were more and more frequent, while the number of instruments imported from France (for example from Ducretet, Carpentier or Koenig) remained important, but was slowly declining.

In the last decade of the nineteenth century, thanks to some newly-founded instrument-making firms, like the above-mentioned Officine Galileo, more Italian instruments arrived in the collection. But the signatures on the instruments are sometime misleading. For example, several pieces signed by the Florentine maker A. Dall'Eco (or by his successor G. Santarelli) were in fact imported and were made by foreign instrument makers such as Elliott Brothers (a moving coil galvanometer), Max Kohl (a 60 cm spark induction coil) or Carpentier (a large Weiss electromagnet). From the old inventories we can see that several other instruments, which are not signed, were acquired from various other Italian retailers. From their design and their technical characteristics, however, it is evident that they also were made in France, Germany or England.

When we examine the old inventories of the physics cabinet, we can also readily see which instruments entered in the collection and when. The collection was not static but was continuously enriched at least until World War One. Every year new instruments and models were purchased and, if we examine the dates of acquisition, it is possible to see how the collection was constantly updated, following the most recent progresses of physics and technology. For example, to the old-fashioned double-barrelled vacuum pumps, which continued to be used for simple didactic demonstrations until the beginning of the twentieth century, were added better-performing mercury pumps (Sprengel and Geissler type) and later rotary oil or mercury pumps of the Gaede type. Certainly, older instruments were repaired and continued to be used, but at the same time other more efficient apparatus was bought.

It is also remarkable to notice that new instruments, which were related to the latest discoveries, were acquired only shortly after their invention. A few examples taken from the section dedicated to electromagnetism show how fast the cabinet was updated. A model of a bi-phase motor proposed by Braun in 1892 was acquired two years later, a Tesla transformer (perfected in the early 1890s) entered in the collection in 1896, several pieces of apparatus of Marconi were acquired in the years 1897–1899, just a short time after Marconi's experiments, a Braun's tube described by his inventor in 1898 was listed in 1903, and an Elster and Geitel apparatus for measuring radioactivity, described around 1900, entered in the collection in 1903 (*Fig. 8*). Several similar examples, demonstrating how

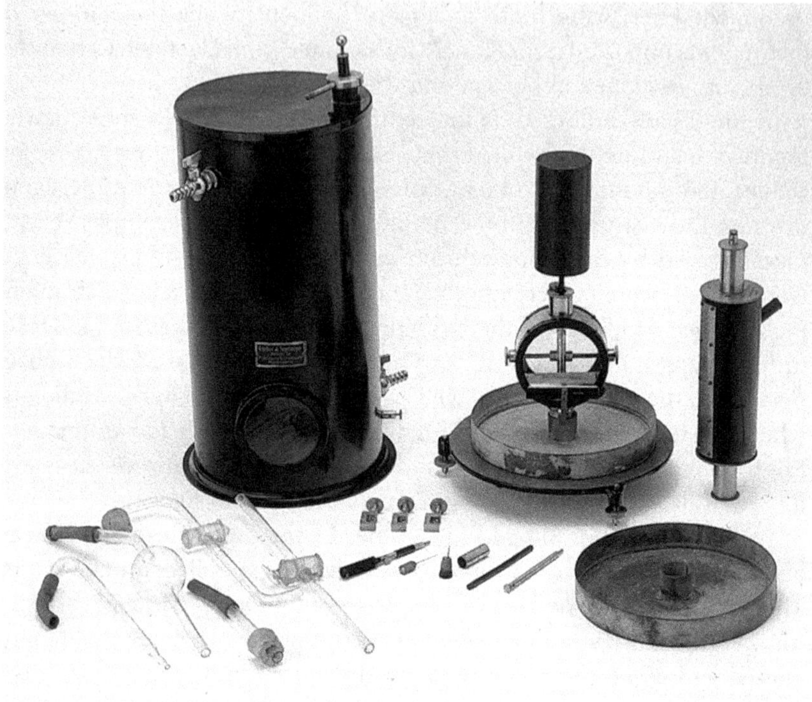

Fig. 8. This apparatus proposed by J. Elster and H.F. Geitel for measuring radioactivity was acquired by the Istituto Tecnico Toscano at the very beginning of the twentieth century (Fondazione Scienza e Tecnica, Florence).

the collection was constantly adjusted, could be taken from sections on optics, heat or acoustics.

This fact is even more remarkable considering that the Istituto was not a research but essentially an educational institution. Such institutions were generally slower to include in their collections apparatus related to the latest progresses in physics. By 1910 the cabinet numbered more than 2,000 instruments and machines, which covered all the fields of classical physics (*Fig. 9*). At the time 2,600 *Lire* were allocated yearly for new acquisitions for the cabinet, while another 2,000 *Lire* were given for its operating costs. Unfortunately we do not have any precise figure for the economic value of the physics collection between 1850 and 1875. However between 1878 and 1915 the increase can be seen in the following table.[25]

[25] The global value of the collection was simply calculated adding the purchase prices of the single instruments. At the time the *Lira* had the same value of the French Franc.

Fig. 9. The main hall of the cabinet of physics, ca. 1900 (Fondazione Scienza e Tecnica, Florence).

Year	Value (*Lire*)
1878	59,138.90
1889	85,511.69
1900	112,903.83
1910	135,233.83
1915	141,948.83

While conversion to a modern value is always difficult and debatable, we can estimate that the sum given for 1915 corresponds approximately to over half a million Euro today.[26] If we consider that not only the physics collection but also those for chemistry, natural history, surveying and agriculture, as well as the technological museum and the library, were constantly being enriched, we can see that the Istituto could count on a very important and constant funding over several decades.[27] In fact by the

26 For the conversion see: www.oppo.it/tabelle/riv-lira-dal1861.html (accessed 2 January 2013).

27 For an overview of the scientific collections of the Istituto preserved today see: Anna Giatti, Stefania Lotti (eds.), *Le stanze della scienza. Le collezioni dell'Istituto Tecnico Toscano a Firenze*, Florence, 2006.

beginning of the twentieth century, the cabinet of physics was certainly one of the best equipped in Italy; its collections were comparable to those of the most important Italian or European universities and polytechnics.

At the end of the nineteenth century, after the relocation of the Istituto to the new building, it was decided to publish a lavish illustrated volume, to be presented at the 1900 Paris universal exhibition.[28] After a historical introduction, a series of articles described the characteristics of the schools, of the laboratories and of the collections. Bazzi, who at the time was the director of the physics cabinet, wrote an enthusiastic description of it. The technical equipment in the new location reflected the state of the art of the 1890s. The collection was displayed in several rooms. Stored in a large hall were the instruments of physical mechanics, hydrostatics, pneumatics and electricity, while in another room were the collections of optics and acoustics. Contiguous to the hall was the lecture amphitheatre with about 60 seats. The mechanical models and the kinematics collection were displayed in a long corridor. Adjacent to it were the office of the director, where some particularly valuable measurement apparatus was stored, and the mechanical workshop.

On the first floor one could find a photographic cabinet, three rooms for the carpenter workshop and the store for the old disused apparatus. In the cellar was installed the technical equipment, which included a dynamo machine and an alternator, both driven by a 5 HP Otto gas engine, and a battery of 60 accumulators, while water, gas and electricity, coming from the city networks, were distributed to all the rooms of the cabinet. Finally Bazzi mentioned some of the most important, expensive or sophisticated instruments in the collection. Furthermore, if not explicitly mentioned in the publication, the cabinet was equipped with a large arc-lamp projector and an impressive collection of lantern slides illustrating apparatus, instruments, modern as well as historical experiments, and portraits of famous scientists. In fact, from the second half of the nineteenth century, didactic projections became common during the lessons and several specialized firms offered in their catalogues thousands of slides concerning art, science, history, contemporary events, etc., which could be bought or rented.

A beautiful photograph taken around 1900 (but not published in the illustrated volume), shows the physics amphitheatre crowded with students

---

[28] See: *L'Istituto tecnico di Firenze: La sua storia ed i suoi gabinetti*, Florence, 1900. An updated edition of the volume appeared in 1910.

Fig. 10. Professor Bazzi and his assistants during a lesson of physics, ca. 1900 (Archivi Alinari-archivio Alinari, Florence).

and full of instruments. In it, professor Bazzi and his assistants show an impressive number of demonstrations: two complete settings for the Tesla's high frequency current experiments (with a Ruhmkorff induction coil and two types of Tesla transformers), a four-prism spectroscope with its spectral lamp, a series of Geissler's and Plucker's tubes, the large Ruhmkorff electromagnet, a projection lantern with its arc lamps, and several other instruments (*Fig. 10*). Certainly this photograph gives an idea of what a lesson could have been. It is possible, however, as is often the case with this kind of image, that the setting was carefully prepared for a posed picture, which was probably intended for a publication.[29] Therefore such a document did not necessarily reproduce the real situation, but was an idealised image of what the reality could have been.

29 A second version of the picture, with fewer instruments and without any people was published in the volume: Ibid., unnumbered plate between p. 48 and p. 49.

Conclusion

The ambitious projects of Corridi could be accomplished only partially. The vision of an institute comprising a school with its collections, an academy and a technological museum, and accompanied by series of exhibitions, which could prepare a new generation of engineers and technicians capable of boosting the industrialization of Tuscany, was perhaps too ambitious and clashed sometimes with a series of other interests and factors. The structure, organisation, chairs and teaching of the Istituto Tecnico constantly evolved with the reforms of the educational system and were strongly influenced by local and national political interests, as well as by the rapid evolution of society. All its transformations were closely interwoven with the changes happening in Tuscany and Italy in the second half of the nineteenth and in the first decades of the twentieth century. Certainly in the first years of its existence the Istituto represented, at least in Italy, a completely unique and innovative experience. In the following decades, it remained one of the most important technical school in Italy and marked several generations of Tuscan students.

As happened for several nineteenth-century collections, most of the instruments of the cabinet began to be obsolete in the interwar period. Because of Gentile's school reform, at the same time physics teaching became less and less important in the curricula of the Istituto. Furthermore, in the second half of the twentieth century not only was there a radical change in teaching (first with the introduction of much more simplified apparatus and didactic films, later with the coming of electronic media) but many experiments and demonstrations, which had been repeated over and over for more than a century, were abandoned in favour of a more modern approach to physics. Most of the instruments accumulated between ca. 1850 and ca. 1915 were slowly abandoned to the dust of their cupboards and the tacit knowledge associated with them faded away. Fortunately, because of a series of lucky circumstances, the physics collection was neither destroyed nor dispersed and survived almost complete until the 1980s, when it was rediscovered and his cultural value was fully recognized. In the following years, thanks to long and patient work, the cabinet of physics was refurbished and all the instruments were restored, reordered and catalogued.[30] Since 1987 the Fondazione Scienza

[30] For the instruments of the physics collection see: Paolo Brenni, *Gli strumenti del Gabinetto di Fisica dell'Istituto Tecnico Toscano. Acustica*, Firenze, 1986; Paolo Brenni, *Gli*

e Tecnica in Florence has been responsible for the preservation and the study of the scientific heritage of the Istituto, as well as for the didactic and research activities related with the collections. Since 2007 the refurbished cabinet has been open to the public and to scholars (*Pl. XIII*). About 3,000 instruments can be seen in the original rooms, which perfectly maintain the characteristics of the late-nineteenth-century setting, and today the scientific heritage of this cabinet of the former Istituto Tecnico Toscano represents an invaluable source of information for the history of nineteenth-century physics and its teaching, as well as for the history of instruments and their industry and trade.

---

strumenti del Gabinetto di Fisica dell'Istituto Tecnico Toscano. Ottica, Firenze, 1995; Paolo Brenni, *Gli strumenti del Gabinetto di Fisica dell'Istituto Tecnico Toscano. Elettricità e magnetismo*, Firenze, 2000; Paolo Brenni, "Catalogo degli strumenti di acustica", in Giatti, Miniati (eds.), *L'acustica…*, op. cit. (n. 8), pp. 73–144; Paolo Brenni, "The Physics collection of the Istituto Tecnico Toscano: 30 Years of Work", in Ewa Wyka, Maciej Kluza, Anna Karolina Zawada (eds.), *East and West: The Common European Heritage. Proceedings of the XXV Scientific Instrument Symposium*, Kraków, 2006, pp. 215–218; Paolo Brenni, *Il Gabinetto di Fisica dell'Istituto Tecnico Toscano. Guida alla visita*, Firenze, 2009.

GENERAL INDEX

www.ingramcontent.com/pod-product-compliance
Lightning Source LLC
LaVergne TN
LVHW010641300926
849161LV00005B/1011
* 9 7 8 9 0 0 4 2 5 2 9 6 7 *